THE ROOTS OF HUMAN SEXUALITY

Review of the Stages Between the Animal Condition and Present Human Culture

M. Maurice Abitbol, M.D.

SPUYTEN DUYVIL
New York City

ISBN 978-1-933132-62-4

Library of Congress Cataloging-in-Publication Data

Abitbol, M. Maurice.
The roots of human sexuality : review of the stages between the animal condition and present human culture / M. Maurice Abitbol.
p. cm.
Includes bibliographical references.
ISBN 978-1-933132-62-4
1. Sex. 2. Sexual behavior in animals. 3. Evolution. I. Title.
[DNLM: 1. Sexuality--history. 2. Evolution. 3. Sexual Behavior--history. 4. Sexual Behavior, Animal. HQ 12 A149r 2010]
HQ21.A25 2010
306.7--dc22

2010019802

TABLE OF CONTENTS

ROLE OF SEXUALITY IN HUMAN EVOLUTION

Introduction

As a practicing obstetrician and gynecologist, I have thoroughly meditated on all aspects of human reproduction. I quickly learned that the function and purpose of human sexuality is not solely directed toward reproduction as in animals. In fact, this is how human sexuality differs from that of animals. I became very critical of a concept that limits human sexuality to reproduction and that establishes a strong similarity between us and animals. I came to the conclusion that sexuality really has been and still is a new force, some kind of engine that has motivated and guided our prehuman and human ancestors to undertake the arduous path of human evolution.

Why has the evolution of sexual behavior been selected in this study as the prototype for the evolution of human behavior? What is sex all about? The entire sexual act consists of penetration of an erect penis into a vaginal cavity, some thrusts, ejaculation, then withdrawal. Later I will show that, unlike humans, the fundamental characteristic of animal sexuality is its casual aspect. The animal female, as far as one can observe, does not consider sexual penetration to be different in any way from any other form of physical contact such as touching each other, grooming, etc. Under specific circumstances, the human female also can be casual in this matter. During my career, I have often counselled women on vaginal douching, insertion of tampons and diaphragms, and I have performed numerous speculum insertions and vaginal examinations. All these manipulations are readily accepted by these women. Yet, the sexual act, which is roughly characterized by similar mechanical gestures, suddenly becomes an act of great significance and enormous implications. The organized grouping of mechanical gestures that composes coitus no doubt has become the most important activity which involves men and women. One forgets to ask why it has become so, and why it has remained so. The steps leading to the present condition are hard to reconstruct but can be found in anthropology, popular cultures, etc.

I believe that the best way to investigate the evolution of behavior in general, and social behavior in particular, is by studying the evolution of sexuality. It is during the sexual act that individuals most closely interact, and this is where prehuman and human behaviors have been primitively manifested. Men and women will be seen completely reorganizing their sexual lives and using it to force this new type of human evolution. Without it *Homo sapiens* might not have made it. A sexual revolution has occurred during the evolution from animals to *Homo sapiens*. Obviously, this transformation did not occur suddenly but progressively, step-by-step. Every known fossil hominid must have had its own specific sexual pattern. In thisi manuscript, an attempt will be made at describing the different "landmarks" leading to present human sexuality.

This undertaking required an exploration of general animal behavior, including their reproductive and sexual behavior, and a thorough comparison with the corresponding human behavior, in order to outline their basic differences. It also required a thorough anthropological investigation of available prehuman fossils to reconstruct the intermediary steps between animal and human sexuality.

I will describe the role sexuality played in human evolution and will only address the current crisis of human sexuality from the evolutionary point of view.

Evolution of Bipedalism

About 3 million years ago a new creature appeared on earth, Al 288-1 or "Lucy." Unlike other apes, this new species presented a specific characteristic that would have important consequences: she stood erect. Any nonhuman primate, the chimpanzee for instance, could stand up and move bipedally. The fundamental difference was that Lucy stood up as a way of life, as evidenced by her low, broad hipbone as opposed to the long and narrow hipbone of nonhuman mammals (Stern and Susman, 1983), the human shape of the sacrum (Abitbol, 1987), and the Laetoli footprints (Leaky, 1964). This study will not involve the current controversy about whether Lucy really stood up (Sarmiento, 1987) or still showed traces of arboreality (Stern and Susman, 1983), or if her bipedalism was as or more perfect than humans (Lovejoy *et al*, 1973), or whether this bipedalism was of the waddling or the striding type.

It is possible that further archeological diggings may unearth creatures that have preceded Lucy in terms of erect posture behavior. It is not even denied that if apres are properly trained, they can be conditioned to use bipedalism almost exclusively at the expense of quadrupedalism. The main point of this study is that a new creature appeared on earth, some kind of animal not generally different from the present chimpanzee, a new ape with the characteristics of bipedal posture/locomotion. These apes are called hominids, and the transformation from nonhuman primates to humans is called hominid evolution. All these hominids are now extinct except for the last one, which is represented by present humans (*Homo sapiens*).

The study of evolution of general behavior from animal-like to the human condition is a difficult and complex task. One reason is that it crosses multiple and dissimilar scientific disciplines such as anthropology, archeology, biology, medicine, animal and human psychology, psychiatry, and philosophy. These disciplines are all necessary for the reconstruction of this evolution.

In a first attempt to deal with the evolution of behavior, authors have investigated in depth the general behavior of apes. Chimpanzees were studied by Jane Goodall (1971), and Diane Fossey (1983) studied the behavior of apes. They in fact lived with these animals for years. Their reports give the impression that isolated elements of human behavior can already be seen in these animals. In a different approach, other researchers (Margaret Mead, 1963) have specialized in what is called "primitive" human society. They studied these societies in the Pacific Islands, Central Africa, and South America to find "remnants" of extinct human societies. Yet another approach is to investigate a few archeological remnants of the now extinct prehuman forms that have been excavated to date, and with major speculations reconstruct these transitional societies. Despite all scientific efforts, the nature of life for these different missing links (Lucy, *Homo habilis, Homo erectus*, the Neandertal, archaic *Homo sapiens*) is still a mystery. How we evolved from animal to present human life is deeply buried in archeological sites, maybe forever inaccessible to investigators.

The deductive technique (scientific discovery based on observed facts) is only one of the methods. More and more discoveries are based on inductive technique (an idea based solely on intuition is improvised and searched for secondary confirmation). Many scientific facts (atomic structure, electromagnetism) are widely accepted today, although scientific confirmation based on direct observation is yet to come. There is no reason not to use a similar technique to reconstruct the evolution of human behavior.

Researchers could distinguish several types of hominid evolution from an ape-like species to present *Homo sapiens*: upright posture, brain size or encephalization, dentition, intelligence (thinking and feeling), and sexuality. The first three have been extensively studied. The evolution of intelligence has been conspicuously speculated upon and attempts at description, based on meager archeological findings, have been made by different authors. Evolution of sexuality, although interesting, hardly has been confronted. It will be attempted in this manuscript.

The Evolution of Sexuality

Are animals right in being casual about sex, and are we wrong in magnifying its significance to such a degree? Has the whole concept of sex become a pathological deviation of our human mind, and have humans become involved in mass hysteria? Is there justification for a purely mechanical approach to sexuality? The emotional reaction associated with sex (or lack of it) is puzzling. Some people feel that the casual approach of animal sexuality has some merit, that, sexually speaking, humans have regressed, and that the animal approach should be copied and envied. This is certainly the opinion of Camille Paglia (1990), the "sexual" philosopher. It is not my intention to take sides on any debate concerning human sexuality, but rather to ask how the casual aspect disappeared and how this new concept of sexuality which played such an important and magnified role in human life evolved and took precedence. At some point in the evolution of our present human culture, not only has the sexual act itself taken on special significance, but any gesture which even remotely may have a sexual connotation has also acquired a special significance.

Today, we live in a bizarre paradox. On one hand, the quantity, quality, and intensity of sexual stimulation is on the rise. One only has to look at daytime soap operas, glance through 'Victoria Secrets' catalog, or listen to some suggestive lyrics. Visual, auditory, tactile, and olfactory (perfumes) stimuli have increased until it seems that we are constantly bathing in an atmosphere of sexual stimulation. On the other hand, we live in an atmosphere where taboos connected to sexuality (the case of Anita Hill versus Chief Justice Clarence Thomas, the William Kennedy Smith case, the much publicized case of President Boris Yeltsin tapping a female executive on the shoulder, and the President Clinton-Monica Lewinsky affair) are quickly distorted, magnified and the focus of society's attention. This new interpretation of human sexuality is, in my opinion, one of the most important events that occurred during human evolution, along with erect posture and increasing intelligence. How sexual activity progressed to such a high level of significance during human evolution is the subject of this study. After all, it is in sexual studies that Sigmund Freud ignited the psychological revolution of the 20th century by demonstrating the reality of sexual dreams, the Oedipus complex, and engaged in the study of incest, etc. One may be justified in asking how it all arose during human evolution.

Misconception of Animal Sexuality

There are many reasons for misconceptions of animal sexuality. The first one is that the study of sexuality, in general, is always difficult because it constantly has been associated with embarrassment, shame, taboos, ordeals, etc. For a long time it has been and still is considered an intimacy, not suitable for scientific inquiries. Freud's studies on sexuality and their influence on our way of life have not completely erased these taboos. Scientific investigation is not the only way to study animal and human behavior; speculation and science fiction often contribute. This is what is happening with the evolution of sexual behavior. In my opinion, animal sexuality has been fantasized about more than any other behavior. When animal behavior, including sexual behavior, is reported by nonscientific observers, pseudoscientists, or simply people writing books or making films on this subject, frequently partiality is introduced in their observations. It is frustrating to observe how science fiction programs such as "Planet of the Apes," "The Intelligent Apes," and the "Speaking Animals" attract a large audience, while programs based on accurate observations such as those of Diane Fossey (1983) and Jane Goodall (1971) draw a relatively limited audience.

The second reason is that there is some guilt feeling associated with mankind slowly and progressively taking over the land, sea, and sky from mammals, fish, and birds. This will be discussed later in more detail, but one can witness the multiplication of societies to protect the rights of animals against scientific research, the fur industry, the cattle industry (raising and consumption of meat and use of leather). Organ transplants from animals to humans, a technique now in its infancy, is considered unfair in some circles. Less progress is made in fighting AIDS because scientists are not completely free to experiment with this disease in primates, the only receptive species to this disease

Today, the concept that animals are living individuals with rights similar to humans is rapidly spreading. These policies are justified and certainly protect animals against human abuse and cruelty. The Swedish model is certainly an example to be followed worldwide. These movements use different kinds of publicity to make the world aware of the dangers facing disappearing species. Animal life, however, is presented to the general public by these organizations in a simplified and idealized manner. The danger is that thinking and feelings similar to humans are easily attributed and projected onto animals and create distortions about real animal life. In this matter, people can overlook the fact that animal sexuality is completely different from human sexuality in every respect, as we will see later.

The third reason may simply come from the fact that men and women presently have tremendous difficulty in defining their position in this world and are vascillating between the simple and attractive animal life and the complex and intimidating higher human values. As stated by Nietzche (1978), "man is walking on a rope stretched between the animal and the superman, a rope over an abyss." Aspiring to high spiritual values (Teilhard de Chardin, 1958) may certainly appear frightening to many human individuals. Commenting on freedom, the highest human acquisition, Dostoevski (1962) said, "Nothing has been more insupportable for a man and a human society than freedom." Carl Jung (1957a) added, "The price (of freedom) is an inner world of fear and insecurity and loneliness." Identification with high spiritual values may be hard to achieve, whereas identification with our animal origins is certainly easier. Animal life is glorified on television programs such as "Wild Life in America" and by others (De Unamino, 1954) who claim that were it not for human interference, animal life would be paradise on earth. The harsh aspects of real animal life in the wild, as will be described, are mostly ignored (Hrdy, 1995).

I will show later that human sexuality is one of the greatest achievements of humans, although it is still subject to many imperfections and errors. Many humans, disappointed in their relations with others and searching in vain for ideal behavior, have turned to animals, and appear to have found the ideal of sex in them. Purity of love, realization of freedom and life without guilt or shame are thought by many to exist only in animals. Today, freedom, harmony, family life, togetherness, solidarity, and open spaces are all concepts accepted as a way of animal life in the wilderness. Sex in animals is considered pure, noncommittal, and free of guilt. On the contrary, sex in humans is thought by many to be complicated, too involved, consequential, disturbing, and frequently associated with guilt. Animals are envied for their interrelationship, love for each other, lifestyle, and casual sex. Since so little is known about their way of life, including their sexual life, animals become an easy receptacle and target for our lack of accomplishments, including sexual ones (see Wildlife in America programs presented by the Public Broadcasting Service).

The final reason is that some animals have become companions and real friends to humans. More and more animals are now living, one way or another, in contact with humans, and this produces tremendous changes in animal behavior. This relationship with animals is to be highly commended and is beneficial for humans as well as animals, since animals play an important role in the emotional life of humans. There is no doubt, however, that this emotional attachment interferes with an open mind which is necessary to remain objective when describing animal behavior in the jungle. The sexual behavior of domesticated animals, as is their general behavior, is entirely regulated by humans, and should never be confused with life in the wilderness. The study of sexuality in domesticated animals, caged animals, and animals living in zoos or artificial habitats, has to take into account the major differences resulting from changes in living conditions. Sexuality in animals living in contact with humans has been altered ("humanized" for better or worse) and could serve as a model for comparison with human sexuality only in a limited way. Whether in the wild or in contact with humans there are, therefore, numerous reasons why humans think animal behavior, especially sexual, is different from the reality of it.

This discussion on the misconception of animal sexuality already gives us an idea of what is and is not animal sexuality. In the following chapters I will explore animal sexuality in depth, and for simplification purposes, this investigation will be carried out in four steps.

First, sexuality in animals and in humans will be studied separately and then compared.

Second and most important, the evolution and acquisition of all sexual characteristics that differentiate humans from animals will be assessed one by one, with a look at why and how such a difference arose, and how it was addressed and eventually solved. At this stage, no attempt will be made to determine which extinct hominid dealt with which problem.

Third, after placing (1) all the fossil hominids in increasing order of encephalization (intelligence) and (2) the sexual problems in an increasing order of complexity, an attempt will be made to determine which sexual problems have been addressed by which fossil hominid.

Fourth and last, generalizations will be made concerning the direction taken by human sexuality.

ANIMAL LIFE IN GENERAL

Introduction

The best approach to the study of animal sexuality and its comparison with human sexuality will be to start with animal behavior in general, then focus on their sexual behavior, and finally compare it to humans.

I will first describe how animals relate to each other in the jungle and the methods they use to control their population. Further chapters will study the controversial and sensitive subject of mother-infant bonding and will introduce the subject of emotion and feeling in animals. How animal life has been influenced and modified by human presence also will be investigated in a subsequent chapter.

The present chapter will describe how animals use the five senses to relate to the outside world and how differently they use them compared to humans.

The Five Senses

Our five senses give us only a subjective view of the universe. Mammals perceive the universe. Although both humans and other mammals use the same five senses to perceive the outside world, they use them in different and even opposite ways. The perception of the outside world by animals is completely different from that of humans because not all species use the five senses of taste, scent, sight, hearing, and touch in the same way. In fact, every mammalian species emphasizes one or more senses in preference to others.

(1) Taste. The sense of taste functions in all terrestrial vertebrates, possibly in all vertebrates. However, reptiles (especially snakes) use it extensively as a mode of communication with their surroundings. They also use sight, but only secondarily. All mammalian animals use taste to perceive their surroundings. These animals constantly lick each other, either among individuals of the same species, or of different species. In nonhuman mammals

tasting is not as important as it is in snakes, but it is more important than it is in humans. The part of the brain involved with taste makes up most of the brain in the primitive terrestrial vertebrates (reptiles). This reptilian or "taste" brain persists in nonhuman and human mammals, but it is proportionally smaller compared with the rest of the brain, and in the human species it is of even smaller size.

(2) Scent. What has been said about taste can be repeated about scent, but on a different level. Scent is highly developed in nonhuman mammals and is used extensively as a mode of communication. Animals also use all the other senses (tasting, seeing, hearing, touching) but much less, and the last three senses are secondary to olfaction and taste. It is through sniffing and licking that animals relate to and identify each other. The olfactory function has unbelievable dimensions in the animal kingdom (Trevathan, 1987). Different individuals of the same species give each other an "olfactory" identity or "olfactory" name. A lion, just by smelling the urine of another lion, even when it is dried up, will know immediately who it is, whether it is a member of his pride or an outsider. An "olfactory" trail left behind by any individual is easily followed by a dog or any other mammalian species. In a pack of hunting dogs, the hierarchy established between the different canine individuals may be based on physical strength, but it is also based on the acuteness of the olfactory function, and the individual with the sharpest nose is immediately recognized as a leader (Jerison, 1973).

The olfactory function is located in the midbrain, surrounding the reptile or primitive brain (Hassler and Stephan, 1967). It is predominant in nonhuman mammals. It is called the neomammalian or limbic brain and is essentially concerned with scent.

In humans, anosmia, or loss of olfaction, can be present from birth or lost subsequently, and in this case it can be temporary or permanent. Although losing the faculty of smelling good food may be a handicap for some people, most of the time it passes unnoticed; anosmia is frequently associated with the "flu" and people are barely aware of it. The role of olfaction is very limited in humans, and olfactory sensitivity can vary tremendously between individuals and is the only sense where one individual may have an olfactory function up to 1,000 times more sensitive than another individual; yet, both can be normally functioning individuals without any pathological condition.

Taste and smell very often are combined in the human species and possibly in animals. For instance, when drinking a cup of coffee and appreciating its aroma, it is difficult to differentiate smell and taste. Both senses are biochemical in nature and both suppose a direct chemical action on the terminals of these two nerves, either on the tongue or on the nasal mucosa. It also is possible that these chemical agents are absorbed in the blood stream through the nasal or lingual mucosa and act directly on the corresponding nerve center of the midbrain.

(3) Sight. Vision is present in all vertebrates and in many non-vertebrate species. Visual perception generally is not as acute in animals as it is in humans, since many animals are color blind and others do not have even a tridimensional visual perception of their surroundings. Its function is not as important in animals (including non-human primates) as it is in humans (Sinclair, 1985; Oyama *et al*, 1979). Although many nocturnal animals have above-normal sight and can see in the dark, in no other animal does this function reach the level of development as in humans. It is not only in refined perception of color and in binocular vision (depth perception) that human vision is superior to most animals. It is mostly in the tremendous development of the visual center in the cerebral hemispheres and the numerous connections between these visual centers and other parts of the brain (Kappers *et al*, 1960). Visual centers and their connections contribute significantly to encephalization (or characteristic hypertrophy of the brain in humans). In other words, it is mostly by what they make of visual sensations that humans are superior to animals. The visual sensations in humans are directed to the cerebral cortex where they are interpreted and then eventually initiate a motor response.

(4) Hearing. What has been said about vision can be repeated about hearing. Animals appear to be sensitive to the slightest noise because they have big and mobile ear auricles that can be directed toward any noise. It is, however, in the higher auditory center of the brain and its connection with the speech center that the auditory system of humans is far superior to that of animals. Not

only can humans perceive sounds of various frequencies, but what humans make of these perceived sounds is different from what other animals make of them. The speech and hearing centers of the brain and the interconnection with different parts of the brain also contribute to encephalization, similar to the visual centers (Tobias, 1971).

There is another important difference between animals and humans with regard to the auditory system. The auditory system of the human fetus is completely formed and functional from the beginning of the third trimester of gestation, and at birth it is no different from that of an adult. The fetus effectively hears and responds to noises (breathing, heart beat, gastrointestinal tract) that may come from the maternal body (Abitbol *et al*, 1992). Strong noises coming from outside the maternal body also can be perceived. Most of all, recent research has demonstrated that the fetus in the third trimester is responsive to its maternal voice. The implications are extensive because of the classic assumption that the neonate comes into the world completely "blank" (tabula rasa) and only from birth can the external world have any influence on the neonate.

Since animals do not use any language, the noises they make have different, if any, meaning. For instance, many animals dispense from making any noise. Except for some monkeys, nonhuman mammals, pregnant or not, barely make any noise (Tinklepaugh and Hartman, 1930; Thorpe and North, 1965).

(5) Touch. What has been said about seeing and hearing also applies to touching. Touching is limited in animals, but this statement needs to be qualified. Any person attempting to domesticate or tame an animal knows the importance of constant petting and caressing (Montagu, 1971). It is true that all animals sense and react to being touched but they rarely use it as a mode of communication or as a way of establishing a relationship the way humans do. Essentially, all nonprimate mammals who have no hands touch only when they bump into each other. There are only a few exceptions: maternal-infant bonding to provide warmth, protection, and nourishment; close aggregation of adults during cold weather or in case of danger; and in a few scattered instances, romping as it occurs between bears, etc. The other important exception is

demonstrated by different primates who continuously indulge in grooming because they have acquired fingers or handlike limbs. Many nonhuman mammals may give the impression of touching the environment for identification purposes, but what they are really doing is using their hands or feet to bring subjects or objects to their mouth and/or nose for identification by tasting/sniffing.

What has been said about the visual and auditory pathways and corresponding areas in the cerebral hemispheres also apply to touching, since its nervous pathways and sensory centers are in the cerebral hemispheres. Compared with other mammals, the areas of the cerebral hemisphere concerned with touch are enlarged, and along with their connection to other parts of the brain, they contribute to encephalization (Norback and Montagna, 1970; Sarnat and Netsky, 1974).

Use of the Five Senses

The differences between animals and humans are basic and involve the nature of the utilized modes of communication as well as the zones of the brain from which they originate. Actually, animals live in a world where they perceive, relate, and communicate in ways that are completely alien to humans. Animals continuously sniff and lick, and humans continuously look, listen, and touch. Only in light of this observation can one clearly understand the difference between animals and humans. The best example, of course, is the dog. Humans want to impose their modes of communication on dogs; they want the dog to establish eye contact, to respond to verbal orders, and to accept their patting. Through training, animals are conditioned to these completely new modes of communication and relationship. Confusion begins when lack of eye contact by the dog is considered lack of any contact, lack of attention to verbal orders is considered inattention, etc. The confusion worsens when the animal cannot stop continuously sniffing and licking the human person with whom the relationship is attempted because the involved human considers it an expression of deep intimacy as it would be between humans. The truth is that when a dog sniffs and licks a human (as it does in its own world), it is equivalent to humans looking at or talking to each other.

Because humans perceive the world and others almost exclusively through vision, hearing, and touch, deprivation of one of these senses (blindness, deafness, or bilateral hand amputation) results in a severe handicap. Scent and taste are accessories. Imagine a group of people in a room, all blindfolded, ears plugged, hands tied behind their backs, and obliged to relate to each other only by sniffing and licking: this gives us an idea (though exaggerated) of what animal life is like. This concept is not too exaggerated, however, because vision, hearing, and especially touch, are not developed in nonhuman mammals the way they are in humans.

Experimental interference with smell and taste in laboratory animals results in a marked disturbance of their normal life as well as sexual life. Interference with sight, touch, and hearing in animals results in some functional disturbances but is not as serious as in humans (Chusid, 1982). In fact, one or more of these three senses may be defective or normally absent in some lower mammalian species, and yet these species manage to function very well in their everyday life, including sexually. A domestic dog, for instance, who is a member of a household, can be trained to function satisfactorily although blindfolded and with ears plugged. I have personally trained dogs who had their ears plugged and eyes blindfolded and who, after a period of training, function sufficiently in a home they identify the environment and they recognize people by relying solely on sniffing and licking.

Pheromones are special hormones produced only by animals (Vandenbergh, 1983; Fox, 1975). They are different from ordinary hormones which react only within the body of the individual producing them. Pheromones also are produced by an individual, but they react on another individual of the same species who perceives them through sniffing and/or licking the individual producing them. They act as an olfactory mode of communication and interaction between individuals. Pheromones produced by animals are found in urine, vaginal secretions, ventral glands, feces, and general body odor. This concept of pheromonal stimulation is presently unknown in humans (Bronson, 1968, 1971). Animal relationships are consequently regulated by biochemistry and we, humans, barely understood until recently what that means (Izard and Vandenbergh, 1982a). Lately, it has been discovered that most animal interactions are biochemically determined and mediated by pheromones that are produced by one individual, sniffed by and reacted upon by other individuals of the same species. Pheromones are immensely important since they regulate mother-offspring bonding, sexual relationship, hierarchy, and many other activities. Our knowledge of pheromone activity is expanding more and more (Izard and Vandenbergh, 1982b). Periodically, one can read popular reports on pheromone activity in humans, namely sexual stimulation through scent and taste (Forsyth, 1966; Ackerman, 1990). These isolated reports are in need of rigorous scientific investigation.

It would be a mistake to think that, in terms of intelligence and emotional maturity, the whole human potential is a function of encephalization (increase in auditory, visual, tactile areas of the brain). Sensory and motor pathways are the mode of expression of the human condition rather than its component. It has been repeatedly demonstrated that blind, deaf, mute people (or any combination thereof) can reach a high level of intellectual and emotional maturity if circumstances are favorable and proper training is available (Hawking, 1988). An important remark is in order here. Blind people develop a heightened sense of hearing and touch, and blind/deaf people develop a tremendous sense of touch, but none of these handicaps are associated with regression to a sniffing/licking obsession considered to be pathological. Only in some pathological conditions, such as certain forms of childhood schizophrenia, do humans revert to sniffing and tasting as ordinary modes of communication (Johnson, et al, 1982; Keverne, 1974; Meredith, 1980, 1982; McClintock, 1978, 1981).

Identification of an individual by voice, sight, and hearing are exclusive human characteristics. Animals, too, look at each other's face and eyes and hear and listen to each other. However, these modes of relationship are not as important in animals as they are in humans. When animals bring their faces close together it is not for eye contact but for the purpose of sniffing and licking each other. Animals rarely communicate vocally, and although vocalization is present, it is secondary and limited to a few specific sounds (birds singing, dogs barking, etc.).

General Traits of Animal Behavior

The way animals use the five senses will give us some idea of their general behavior. We will start by describing certain traits that characterize the great mammals (nonhuman mammals). Common observation shows that general animal behavior rests on a few basic principles that have been recently popularized by Robert Ardrey (1961, 1966), among others (Altmann, 1962; DeVore, 1965).

Territorial imperative. One individual, or a group of individuals of the same species, delineates a territory as its own with rigid frontiers usually marked by scent (urine, feces, etc.). Battles to the death are common occurrences in defending territorial limits.

Gregarious instinct. Animals of the same species tend to aggregate, and the size of the grouping varies. It can be thousands of individuals (herbivores or banks of fish), or limited to a few individuals such as one male and many females like lions (Caras, 1985; Bertram, 1978), or individuals like tigers (Dominis and Edey, 1968; Mountfort, 1981) who, except for mating, live alone. The reasons animals aggregate are obscure, and the explanations for this basic instinct are multiple and complex. It is possible the similar olfactory identity creates a biological attraction, and also that the togetherness creates an atmosphere of security. This may be a reason animals form schools, flocks, herds, and all sorts of social groups. This aggregation of individuals takes different forms depending on the species.

Hierarchy. A system of hierarchy is an absolute law for animals living together and usually is based on physical strength. The individual at the top establishes dominance over the others and has absolute and total privileges including food, females, choice location for sleeping, etc. The second individual from the top has the same but lesser privileges, and so on, down to the individual at the bottom. There are continuous confrontations, very often deadly, as the hierarchy is constantly challenged and reorganized. There is no democracy in animal society and individuals have no (unearned) rights. This is completely different from human society where, even in the most primitive and authoritarian forms, some aspect of individual rights is always respected.

Cerebral center. There is a preponderance of the midbrain over the neocortex. All animals have biological needs (hunger, thirst, sleep, sexual, defecation, urination, etc.) with the neurological centers of all these biological manifestations located in the midbrain. The neocortex or cerebral cortex (cortical brain) contains the motor and sensory center (sight, touch, hearing) and is also the center of intelligence. It is limited in animals (Armstrong and Falk, 1982). The important characteristic of animal behavior is that when a biological order (hunger, for instance) issues from the midbrain, there is little, if anything, that the cerebral cortex can do about it except fulfill it, and if not carried out immediately, the animal becomes enraged, making it dangerous to others. The cortical intelligence in animals is not capable of inhibiting, modifying, or initiating any behavior. If the animal is not hungry, it will not look for food; if the male animal is not stimulated sexually, it will not look for a female. In other words, the cortex is subordinate to the midbrain (Chusid, 1982).

In humans the cortex functions, or rather can function, differently. Here, too, the midbrain initiates the same kind of orders as in animals (hunger, thirst, sleep, sex, etc.), but the human cortex is capable of exerting enormous control (almost total according to Kant, 1956) over the midbrain orders. The human cortex can inhibit, modify, or delay these midbrain orders. More importantly, the human cortex can initiate its own orders independently of the midbrain. For instance, *Homo sapiens* can fast when hungry or eat when not hungry, while animals must eat if hungry and will not eat if not hungry. The only exception is when the animal has been trained by humans to modify its behavior. This conditioned reflex (Pavlov) will be discussed later. In my opinion, the major difference between animals and humans is that the animal midbrain controls the cortex, while in humans the cortex controls the midbrain. It is fair to say, however, that human control of the cortex over the midbrain is sometimes more potential than real, and regression to something similar to an animal condition, where the instincts of the midbrain are not fully controlled by the cortex, is not uncommon (William Golding, 1962).

Lack of activity. Quoting Adrian Forsyth (1966): "Most animals seem appallingly lazy. For most of the day they lie about dozing, sunning, and generally goofing off when they could be busy looking for food or otherwise trying to get ahead in life. Anyone who gets his impressions about wildlife from television specials is in for a shock of boredom if he ever attempts nature photography or animal watching. For every shot of a group of lions pulling down a wildebeest, the camera crew spends hours, sometimes days, watching the lions loll about ...[Study] in animal behavior may mean spending hundreds, even thousands, of hours watching animals sleep and scratch themselves, and only a small amount of time seeing them do interesting things...There is no point in hunting when your stomach is full or when there is no food to be had."

When animals live in artificial enclosures such as zoos or jungle habitats where food is abundantly provided, their indolence is, if anything, greater. The most frequent argument used by those criticizing these man-made enclosures and, in fact, the general impression perceived by visitors, is that these enclosures (and man) are responsible for the animal's continuous boredom, while it is really their normal way of life. As astutely noticed by Forsyth, animals do not try "to get ahead in life." This last observation helps to understand the biological and philosophical differences between animals and humans. Living creatures on earth can be divided into three groups: plants, animals, and humans.

Plants are stagnant and rooted to earth. If they survive, it is because all the elements of life (sun, water, nutrients) come to them, since they are barely capable of moving and looking for them. This is why plants have no use for a nervous system.

Animals, like plants, also would have a tendency to be indolent and remain stagnant except that they have to move to search for and find food. For that purpose, animals have developed biological drives, which essentially are biochemical variations determining their behavior. When the main biological drives (food, sex) are not activated, the animal is passive and stagnant, almost like a plant. There are other drives that can make animals move and perform actions such as scratching, yawning, romping, grooming, etc. Basically, however, the animal is stagnant and plant-like and deviates from this condition only under the influence of biological drives.

When we come to humans, the condition changes dramatically. Human behavior still can be determined by biological drives, such as hunger, sexual stimulation by hormones, etc. But humans have a tendency to develop a behavior independent of biological drives (Buber, 1970). A human being can thus be defined as an individual struggling between biological drives and independent behavior. This point has been stressed by philosophers of all ages (McNiff, 1962), from the ancient ones such as Socrates and Plato to the most modern such as Kant (Paton, 1970). To see a change occur in the behavior of an animal species, one has to wait for a genetic mutation. This is not necessary in the human condition.

Genes. There is a preponderance of genetic transmission over learning in animals versus humans. Animals are born with a fixed behavioral pattern that is essentially determined genetically with limited or no difference from one individual to another. The animal mother is necessary for providing three basic needs: food, protection, and warmth (Trevathan, 1987); but any artificial system providing these will produce an offspring grossly similar to one receiving maternal care. The offspring learns very little, if anything (with limited exception for some primates). The bulk of animal behavior is genetically determined and the mother adds very little (Gould, 1977). I know I am on shaky ground when making such a statement, but I would like to emphasize that when observers report behavioral differences in offspring of the same species, they are reporting the result of various human interferences. This will be discussed later. Animals living in contact with humans may display differences that reflect variations in the humans they live with, but in jungle life there is no human interference and the mother and offspring of a given species all have a genetically determined identical behavior. Animal maternal teachings, frequently reported in the general literature and less often in scientific literature (Smuts, 1987; Ghiglieri, 1987), are isolated observations. The animal offspring may observe and copy the mother and thus learn by trial and error, but there is no conscious teaching in the human sense. On the contrary, a human infant isolated from maternal care or other human contact will grow like an animal. There will be no language, little or no sphincter control, delayed posture and locomotion, and essentially predominant midbrain instincts unable to be controlled by the cortex. This dire outcome will occur even though the basic needs (food, warmth, protection) are provided (Curtiss, 1977; Rigler, 1972; Rymer, 1992).

From these six major traits of animal behavior (territorial imperative, gregarious instinct, hierarchy, cerebral center, inactivity, genes), human behavior will evolve.

Animal Life in the Jungle

Lately, there is a general tendency to idealize animal life filled with virtues that are lacking in humans. While some of this is partly true, other aspects of animal life that are not so attractive must be mentioned.

Zoologists, animal psychologists, and students of animal behavior are constantly aware of the fact that it is difficult to objectively study the behavior of animals. Many students have spent years studying animal behavior in the jungle only to find out that their presence has interfered in many ways with the natural animal life. Jane Goodall (1971), for instance, declared that if she had to do her research on chimpanzees all over, she would do her observations from a distance rather than intermingle and interfere with the life of these animals. This in no way diminishes the value of the observations she reported, but there would have been, I imagine, more objectivity. Only films obtained with powerful zoom-lens cameras of animals that do not suspect a human presence can be fully objective.

The overall picture of animal life is rather grim ("bewilderment and terror," as noted by Martin Fisher, 1943). The general impression is one of aggression against other species and within one's own species through hierarchy or territory. The strongest, cruelest, most aggressive is the one who makes it. All this has been well described by Darwin and nothing much has been added since. He mentioned "the clumsy, wasteful, blundering law, and horribly cruel works of nature." We have described purposeful aggression (for food, sex, hierarchy, territory, etc.), but it is my impression that all too often animals are aggressive without purpose.

Basically, animals live in a hostile world surrounded by enemies trying to kill or overpower them. Only one century after

Darwin described it so well, there is a tendency to forget the cruel aspect of natural selection and survival of the fittest. It is wrong to think of natural selection as some kind of olympic game or sport competition, with the audience clapping hands, and medals given to the winner. In the animal world there is a stereotyped "jungle personality," a "fight-or-flight" or "kill-or-get-killed" attitude, with no mercy for the losers: "*vae victis*." Modern technology has made it obvious that the vast majority of animals die of starvation and live in terror of being savagely killed and eaten, or being crushed by hierarchy.

One should never be too quick to think that animals are courageous and always devoted to their family and social group. Frequently, cowardice in animals, like aggression, is a way of life. Predators always attack the weak, and having selected the weak in a herd (old, young, sick), they attack the weakest of them. They often engage in fights-to-the-death, but as soon as one perceives that it is the weaker one, it will run away immediately, giving up females, litter, territory, the prey just killed, etc. A sea lion, or a lion, may own a large pride of females, but as soon as an obviously stronger male shows up, the first owner gives up immediately, and the females will never come to his rescue. I see no evidence of an animal sacrificing itself or even disturbing its way of living in the slightest degree for another individual. An individual will fight, but only if there is a chance of victory. The only exceptions are the often mentioned mother-infant bonding and in animals living in human contact.

The impression of togetherness, so often reported by others, is all too often an illusion (Edey, 1972; Ardrey, 1966). Many authors have described instances of sacrifice by the mother for her offspring, by the leader of an animal group for the safety of the others, for protection of the sick and wounded. Scientific evidence of this is lacking. The few programs I have watched on reliable television channels (Scientific Channel, Discovery Channel, National Geographic, etc.) easily can be explained in terms of biological determinants rather than unselfish sacrifice. For instance, the technique used by wolves running after a deer has been analyzed by "military" experts (Allen and Mech, 1963; Baterman, 1984; Robbins, 1986). The truth is that each wolf takes a position determined by rank: the most dominant chooses the position which will be nearest to the kill, and all the other wolves position themselves accordingly. As soon as the deer is killed, the dominant wolf eats first, the others wait patiently until he has finished, and nobody dares to approach the prey until his turn has come even though two non-dominant individuals acting together could easily displace the dominant one. The skill attributed to predators in attacking a prey is often exaggerated. Certainly, the lion is capable of a fast and powerful pursuit, but only for a short time, and it is pitiful to see how often a lion, or a group of lions, fails in attacking relatively easy prey.

On the other side, it is embarrassing to see the lack of togetherness shown by the predated. Television programs on scientific channels which frequently report on animal life demonstrate this point all the time. A strong and healthy deer can tear apart a wolf, and two ordinary deers acting together could disembowel the strongest wolf. If a few of the strongest deer organized themselves, they could challenge any pack of wolves. Yet, in a herd composed of thousands of individuals it is everyone for himself; the message is clear: "run away." In a specific species, when an individual is stronger than the others, he is really stronger than all the others put together. Hierarchy becomes possible only because there is no togetherness. This complete lack of togetherness makes a single wolf stronger than a herd composed of thousands of deer.

Even the animal mother does not always protect her offspring from the wolves. Careful analysis shows a young one desperately hanging on the mother, sometimes even using the mother's body as a shield against the wolves. The mother is confused and attacks the wolves because she thinks she is protecting herself. As soon as the wolves separate the young one from the mother, they run after the offspring only and the mother no longer interferes and appears indifferent. There are numerous visual reports of such scenes, but I have yet to see a single one showing the mother coming to the rescue of her offspring, once they have been separated.

Survival of the fittest means taking care of oneself first. When an animal is searching for food, struggling for survival, competing for females and territory, or dreading being killed or starving, this animal will be quickly eliminated if it stops to help another individual in distress. Observers investigating animal life attract a large audience when they report love and togetherness among animals, a tendency which is often abused although at times possibly true.

Intelligence in wildlife is of limited use. Take for instance the shark and the dolphin: in the ocean, the shark is the stupidest and the dolphin is the most intelligent. The possibility for love and intelligence in the dolphin is great (Ellis, 1982), while almost nil for the shark. In terms of I.Q., there is more distance between the shark and the dolphin than there is between the dolphin and man. Yet, in the ocean jungle, the dolphin is no match for the shark because the shark is stronger, more ferocious, and that is all that counts.

Similarly, in the forest jungle, the average monkey is more intelligent than the tiger, and yet monkeys become a constant prey for the tigers. A jungle animal is better off with sharper teeth or claws, or longer limbs to run faster, than with a higher I.Q. or more gentleness. Intelligence and emotion are not too useful in the animal world. These qualities in animals are useful only when in contact with humans.

Animals can give the impression of tolerating, playing, or even loving each other, and certain types of behavior reported in different species would certainly give that impression. On closer observation, this is not always true. A relationship established between two animal individuals is always temporary, rudimentary, and casual. Two polar bears may romp together; should competition for food or females arise, their friendship is over and they tear each other apart. When two monkeys groom each other, they may be doing it to satisfy their gregarious or elementary touch instinct and not exclusively as a social activity; besides it is usually not done on an equal basis since the dominant individuals always are groomed by lower individuals in the hierarchy, without much reciprocity (Altmann, 1962).

In animal society, baboons have been the most studied. They have become popular since the Japanese opened numerous baboon parks where these animals live freely. Especially admired in this society is the hierarchal structure. A feature often mentioned is that when a dominant baboon approaches a non-dominant one, the less dominant presents its rump and takes some kind of bending and submissive posture with regard to the dominant one. This has been interpreted as some kind of respectful reverence, as if the baboon had polite manners. Careful observation would have shown that this posture is identical to the one taken by a female as a form of sexual submissiveness, a posture very often taken by terrified females when trying to calm an angry male by proposing sex (Dollard et al, 1961). This respectful posture is nothing but sexual submission on the part of a lower ranking female, or a homosexual submission (Washburn and DeVore, 1961) on the part of a low-ranking male (homosexuality is not uncommon among monkeys), both of them being terrified by the dominant one.

It is also a misconception to imagine that there is division of work in a social unit of baboons or any other animal species (Edey, 1972). The dominant individual does not even bother hunting for prey or searching for food. He just waits for a lower ranking individual to find something and then takes over without any confrontation. High-ranking individuals in the animal hierarchy usually live that way.

The distressing aspect of this hierarchy is that it is considered normal by the lower individuals without any hope for "democracy." Animal behavior will be understood more clearly when humans stop looking at it through human eyes. The mistake often made is that each pattern of behavior in animals is interpreted as if a human being were performing it, and we attribute to the animal the intelligence and emotions of humans performing the same behavior. The behavior of an animal female who helps a young one given up by the real mother should be viewed in terms of satisfying a biological need or olfactory attraction in addition to a noble desire to help a lonely orphan. I am not stating that pure love and intelligence do not exist in animals, it is just that the bulk of animal behavior can be explained by biological determinants.

Population Control in the Animal World

Of all living species, humans have the lowest birth rate and yet they are spreading the fastest. Only humans are capable of controlling their multiplication. Animals have no way of controlling multiplication as it is regulated biologically. If all newborn animals were to live to full adulthood and multiply, within a few generations each and every isolated species would cover the earth with millions and millions of individuals. Natural population control in the animal world is not done before birth (animals in the wild continuously get pregnant and give birth), it is done after birth. The average nonhuman mammalian female gives birth to between ten and one hundred offspring within her reproductive life. How can this incredible number of new individuals not reach adult life? Ninety to 98% of these newborns must die before they reach full maturity and before they can proliferate. The classic explanation is that many of these newborns die at or immediately after birth. This is an obvious extension from humans where perinatal mortality is (or was) rather high (Assali, 1968). This explanation applies only partly to animals because usually the animal infant is very mature at birth, more so than the human newborn, or it matures quickly. Also, most if not all animal litters are well fed by the mother. Perinatal mortality is, therefore, low in animals, certainly lower than in humans. I consider the 30% average perinatal mortality rate in animals, usually quoted in the literature, too high. Assuming this figure is correct, a 60-69% mortality on top of the 30% perinatal mortality has to be accounted for. This is a difficult and sensitive subject to discuss because it supposes the destruction of the vast majority of the animal population in the jungle by cruel techniques. Among these techniques, seven outstanding ones are summarized here.

(1) Predation of the offspring. It is the offspring who are most often hunted. They are the easiest to catch because they are young, defenseless, and do not run fast. Predators usually have a tremendous olfactory function and easily can locate a hidden litter. All a predator has to do is wait for the animal mother to temporarily walk away from the litter in search of food, and when she comes back part of or the whole litter is gone. Many species live solely on predating young ones, and this can be done either within the species itself or between species. Chimpanzees, for instance, practice it not only against other species, but even within their own species (Goodall, 1971).

(2) Rejection of offspring during childhood. When living in

a human home, a close relationship between a dog or a cat and its litter is observed, giving us a false idea about the real relationship between mother and offspring as it occurs in wildlife. Essentially, the relationship between animal mother and her offspring fulfill certain biological functions; one is evacuation of the distended mammary glands, for instance. When a biological change such as cessation of lactation occurs and the function ceases, the offspring may be rejected. The outcome of the rejection is grim: they may starve to death or be predated.

(3) The unprepared adolescent. In a good percentage of cases the young ones manage to attain adolescence (puberty). At the earliest sign of puberty, "maternal" separation occurs almost immediately. Unfortunately, the average adolescent left on its own is unprepared for the struggle. The adolescent predators are not proficient in hunting and the predated are not proficient in running fast or avoiding danger. Competition is fierce, enemies are numerous, danger is everywhere. This is the most dangerous episode of animal life, and it has been estimated that about half the adolescent population does not reach adulthood (Aries, 1982).

(4) Aggregation, hierarchy, and territory. In animal society, individuals of the same species tend to aggregate, form a hierarchy, and establish a territory. These different factors interfere in population control very simply. The dominant individuals organize safety to their advantage by putting themselves in the center of the group (the safest place), where they are surrounded by several layers of progressively less dominant individuals. Predators always attack the group at the periphery because it is the closest, a surprise attack works better on the periphery (the center gets enough warning), and it is made of the weakest individuals who are the easiest to catch. An example would be a herd of deers who are organized the same way, as the dominant ones in the center are spared from the wolves.

Animal population is explosive in nature and food supply diminishes rapidly especially during the winter season so that many individuals have to be disposed ot. Some astute zoologists (Ardrey, 1961) have observed that the herd stick to predators the way predators stick to the herd, with the dominant individuals from the herd "pushing" the weak, sick, old and even the young toward the predators. In relation to a herd of deer, the wolves fulfill the function of scavengers by cleaning the herd of all dispensable individuals or "les bouches inutiles"; the wolves definitely fulfill a necessary function of population control.

The predators not only control the population of the predated, but they control their own population through the "territorial" function. Why do the lion, virtually unchallenged in the jungle of Africa, and the tiger, virtually unchallenged in the jungle of Asia, not have a larger population in view of the abundance of food? The answer is very simple, it is a question of territory. Each individual lion or tiger needs so many square miles just for himself, even if that area can feed more than one lion or tiger. This surface area has more to do with how far these big predators can sniff each other than how many individuals can be fed. Therefore, lions or tigers keep tearing each other apart to keep enough territory for each individual, and this results in their numbers being constantly low. The "territorial imperative" (Ardrey, 1966) is an absolute law in wildlife and contributes more to population control for all jungle inhabitants than does food supply. This population control is exerted not only on land, but also in the air, and in the ocean. Marine mammals, such as dolphins and sea lions, count on the shark to limit their population and keep the species healthy by eliminating the weak, sick, and old. Sharks are ruthless but necessary "friends" of marine mammals.

(5) Mating techniques. Hierarchy excludes a large segment of the male population and some of the female population from the reproductive function. This policy results in a large segment of the animal population being excluded from multiplying.

(6) Short life span. The life span in the jungle is short. Competition among animals is so fierce that any individual not at its top form is ruthlessly eliminated. The two most common forms of death in the jungle are starvation for a predator who cannot kill any more, or falling prey to predators because of reduced speed or weakness. There is no place for sickness or old age in wildlife. This short life span contributes significantly to a reduction in the general population of each species.

(7) Ecological factors. Unexpected weather changes, changes in food supply, shifts in population due to appearance or disap-

pearance of species or their variation in size all contribute to the destruction of jungle animals. Infections, parasites, epidemics, various untreated traumas and other diseases, the most important being dental problems, also play a role in their elimination. It is very important to stress the fact that there is never enough food for animals in the wild. No matter how abundant food is, animals multiply so rapidly that they quickly reach a level compatible with, or above, food supply. Even small increases in animal population can produce frequent episodes of starvation. There is always competition for food, and food is always taken by one individual at the expense of another.

(8) Human interference. Last, but not least, man is responsible for endangered species. Animal territory is being taken over by man because of his expanding population. Laboratories, zoos, jungle habitats, and even the supply of home pets, are contributing to the shrinking animal population in wildlife. For some species, human interference has become predominant and may lead to extinction.

Very few animals know how to save food: a wolf kills a deer, eats voraciously until he is no longer hungry, then forgets about the leftover. He does not think of food until he is hungry again. If there is no kill for him to eat when he is hungry again, the animal will go hungry for a long time and easily starve to death.

It is too early to talk about the present concern of saving endangered species rather than controlling their population. Yet it could have been predicted, and it already has been observed that the creation of national parks, where the species are protected against starvation and predation, is creating overpopulation in some species and dangerous ecological imbalances (Diamond, 1992).

My personal opinions, drawn from the above observations, are as follows. Hierarchy is the opposite of democracy, at least when hierarchy is based on physical strength. When hierarchy (based on physical strength) appears in human societies, it is a remnant of our animal ancestry. Overpowering of females by males, when there is a difference in physical strength between the two genders as it exists in the human species, is animal in origin. Only when hierarchy based on physical strength started to lose ground was human evolution possible. Finally, this reinforces my concept that women, who had more to gain when physical strength lost its hierarchial value, were the original leaders of human evolution.

MOTHER-INFANT BONDING IN ANIMALS

Introduction

Human sexuality involves varying degrees of emotions: love, hate, jealousy, tolerance, faithfulness, aggression and its control, etc. How much these basic emotions exist in animals, and how much they have evolved in *Homo sapiens* is, indeed, difficult to appreciate. The best way to start is to investigate what is known about emotions and feelings in animals, especially the relationship between the animal mother and her offspring.

How much of this animal relationship is purely biological (responsiveness only to biochemical changes) and how much is purely emotional (deep attachment one individual can have for another) will first be investigated.

Nature of the Problem

Up to the 1960s, animal behaviorists took for granted that mothers were instinctively nurturing. But, recent studies drew different conclusions. According to Creel and Worthman (1995) not all females would be mothers. Hrdy and Carter (1995) conclude that motherhood is not just licking, suckling, tending, and awe-inspiring protectiveness for which mother mammals are so justly famous.

My opinion is that in animals, biology determines the nourishing of the progeny in utero as well as after delivery. In humans, biology still determines nourishing of the progeny in utero (with some reservations) but once the infant is born, biology is less important and new human values (emotion, love) step in.

Recent research by authorities in this matter (Schiebinger, 1995 and Badinter, 1981 whose books are the first to question the concept of "maternal instinct;" Hrdy, 1994; and Clutton-Brock, 1991) is revealing. The December 1995 issue of Natural History is devoted to reconsideration of mother-infant bonding in animals as well as humans. Motherhood is no longer a safe topic.

A mother dog may abandon some of her infants and nurture and save the rest. A mother mouse may kill her young after giving birth at one time or be a model mother another time. There is no universal natural response, and there never has been. The mother must weigh the babies at hand against her own well being.

Mongoose and marmosets use "helpers;" otherwise they might abandon their young half the time. In dwarf mongoose and marmosets, a dominant female destroys the pups of any rival.

A golden hamster can easily eat a few of her pups. According to Day and Galef (1995), cannibalizing pups is an "organized part of normal maternal behavior." Biologist John Hoogland (1995) reported that prairie dogs in South Dakota rear most of their litter, but abandon some to be eaten by other dogs or the mother dog herself.

Mohnot and Sommer (1995) and Hrdy and Carter (1995) observed that destruction of young by dominant males is current in Langur monkeys and many other species.

Hrdy and Carter (1995) state that infanticide, abortion and cannibalism are natural lapses from imagined "natural laws."

Today, the animal mother-offspring relationship is cited in the general literature, whether in ants or higher primates, mostly to describe the purity, the intensity, the almost divine aspect of it (Klaus and Kennell, 1982; Newton, 1971). It is practically a heresy today to doubt such reports, and very few authors do (Harlow et al, 1963). The closeness between an animal and her offspring, indeed, can be puzzling. The concept of the pure love an animal mother has for her litter is thought to be so perfect that several authors have suggested it as a model to inspire and guide all human mothers (Trevathan, 1987). It is an accepted fact among some animal psychologists that maternal love is animal in origin, and by studying it we will better understand the relationship between the human mother and her child. I do not doubt the veracity of these reports. I just try to make the description more complete to determine, or attempt to determine, the causes.

Let us observe a baboon colony in a jungle habitat and study the behavior of the little baboons with their mothers. It is impressive to see the baboon infant on its mother's body, how the mother is only a short distance away, how the mother baboon generously makes her body available for protection and refuge, and how she allows the infant to nurse freely from her mammary glands. The mother baboon deserves the highest praise as a mother.

On closer scrutiny, however, one can see that it is the infant who is constantly hanging on the mother, the infant who is terrified when it is only a few feet away from the mother, and only the infant who finds relief when back with the mother. It is basically the infant approaching the mother, and not the opposite. Occasionally, one can see a baboon mother running after her infant and placing it on her back, but we never sense any anxiety on the mother's part because of that short separation. In addition, we never see a mother baboon offering her nipples to her infant for nursing, in spite of the fact that the mother baboon has four very clever hands (not two like humans), which she uses for seeking food and for grooming another baboon. The infant has to find the mother's nipple on its own and has to look for any other food on its own. Visitors may throw peanuts to these animals, but at no time does a mother baboon, or any other adult baboon for that matter, ever pick up a peanut and offer it to a small baboon. Only when a peanut falls to the side unnoticed by an adult baboon, do we see a little baboon leave the mother to quickly grasp the peanut, then rush back to its mother.

The animal mother is very defensive and protective of her offspring, however, and zookeepers know very well that it is dangerous to manipulate the infant in front of the mother. Poachers know that the only way to separate an infant gorilla from its mother is by killing the mother (Fossey, 1983). These observations can be extended to a cow with her calves, a ewe with her lambs, a cat with her kittens, a bitch with her pups. When observing television reports on lions, wolves, deer, and other animals in the wild, one can come to similar conclusions.

Now let us observe a group of human mothers in a park with children of different ages. After careful scrutiny, a fundamental difference in the mother-child relationship between baboon mother and human mother becomes obvious. The behavior of the human mother is strikingly different. The one-year-old human can find comfort in its mother's lap, but it is not scared and has no fear whatsoever walking away from her. Occasionally we see a child crying for its mother if the child becomes lost. Most often we see a mother running after her child, constantly calling it back if it strays too far, and keeping a watchful eye. In fact, the feeling is that not only does the human infant not mind detaching itself from the mother, but at times even seems to enjoy it. When it comes to feeding, or when the infant cries with hunger, the mother always places the child on her lap and guides the nipple of the milk bottle to the infant's mouth. Occasionally, we see a mother offering her breast, guiding her nipple to the infant's mouth and squeezing her breast to make the infant's suckling easier. In fact, most of the time the mother offers food to the child, rather than the child asking for it.

These preliminary observations show that in the mother- offspring relationship there are basic differences between animals and humans. In this chapter the different factors governing the mother-child relationship in animals and the nature of the maternal love expressed for her offspring will be studied.

Conditions of the Mammalian Infant at Birth

The condition of the fetus at birth among different animals is a very important factor in determining the type of care that the newborn needs. In this regard, mammals are divided into three categories.

(1) Altricial mammals, mostly carnivores (tiger, lion, wolf, cat, dog). At birth, the litter is rather large and the creatures are tiny because they weigh only 2-3% of the mother's weight (Rubin, 1963; Altman, 1980). They are helpless, but they do live in a closed and relatively safe environment, since the mother is usually a predator and very capable of defending her litter. At birth, the little ones have closed eyes, are deaf, and barely able to move. The litter is born in a precise spot where they remain until they can move around on their own. The mother lies down beside the litter, and the little ones must find the mother's nipple on their own. The important point is that the mother does nothing to ensure that each individual of the litter is nourished adequately (the way some birds do). Some of these mothers (the cat or the tiger), will carry the litter, by gently grabbing them by the neck, from place to place. After a few weeks, the litter is perfectly capable of moving around

and at that point the relationship between the mother and the litter is similar to the next category.

(2) Precocial mammals (ungulates and primates). They normally give birth to only one offspring usually of larger size compared with the mother's size, about 5% of the mother's weight. The important difference from the altricial mammal is that this offspring is more mature at the time of birth: it can stand on its feet immediately and run after the mother as ungulates do or has enough musculature and coordination in its four limbs to hang onto the mother's body as primates do. In either case, the newborn must constantly seek the mother, run after her, hang onto her, and look for her nipple. Essentially, the maternal role consists of licking the neonate, allowing nursing, and being very protective. Primate mothers like monkeys, chimpanzees, gorillas, orangutans are often seen carrying the infant in the arms, holding it by the hand, or carrying it on their back. In this regard they are somehow similar to humans, although they do not hold their infants as often as human mothers. The rest of their behavior is similar to animals.

(3) The human mother is a paradox because she is both altricial and precocial. She is altricial because she gives birth to the most helpless creature in the mammalian world (especially compared to other primates), and this creature stays helpless for many years. Yet, she is precocial because she gives birth to the largest baby in relation to her own weight (about 7%) which is almost 2-3 times more than other primates. As stated by Martin (1983), the human mother is primarily precocial and secondarily altricial. Different researchers (Trevathan, 1987) have calculated that it is only at one year of age that the human infant reaches the maturity of a newborn ape. The human maternal role is more involved and goes beyond the elementary functions of the animal mother, which are limited to providing food, protection, and warmth (Brazelton, 1979).

Biological Mother-Infant Bonding in Animals

The five biological factors that are activated during the mother-offspring relationship in different animal species are most importantly lactation, then touch, scent, taste, and gregarious instinct.

After the animal has given birth, prolactin is produced in great quantity by the pituitary gland. The mammary glands, which have been hypertrophied during pregnancy under the influence of estrogen and progesterone, are now stimulated by prolactin and become engorged with milk. Milk engorgement in animals can be very painful, much more painful than the breast engorgement experienced by the human species. The pain and discomfort experienced by a cow whose daily milking is delayed gives a better idea of what mammary gland engorgement is in animals. Nipple suckling becomes necessary to empty the distended mammary glands and relieve the painful distention. An animal mother who is abruptly separated from her offspring is certainly suffering from the separation, but the physical pain is due, partly at least, to the significant distention of the mammary glands. The best proof is that when animal mother and infant are separated, administration of hormones relieves breast distention and quiets the mother, although she remains partly agitated due to the separation.

The problem with suckling is complex (Stratton, 1982). Through a neurological pathway starting with nipple stimulation by suckling, the pituitary gland is stimulated to release more prolactin and, therefore, more milk production by the mammary gland. The animal mother is caught in this "vicious cycle" and has no way out but to continuously provide suckling for her offspring. Inhibition of follicle-stimulating hormones (FSH) and estrus by prolactin is crucial for the maintenance of lactation. The important consequence is that the female has no sexual concern during lactation and can devote herself entirely to feeding the offspring. Should prolactin secretion diminish for any reason, milk production will disappear, the need for mammary decongestion will disappear too, and the offspring is in great danger of being rejected by the mother. Mammary congestion and the need for relief are not the only factors determining mother-infant bonding in animals; far from it. But they do play an important role, at least in the beginning.

Observers have a tendency to think that maternal behavior in animals is primarily concerned with feeding the young ones, since this is a human concern. This is certainly true, but bizarre observations can be made that cast some doubts about such an assertion. For instance, the animal mother is not concerned when the stronger ones displace the weaker ones who may starve to death. It is a mistake to think that the animal mother allows the weakest ones to die so the best ones can survive. This teleological thinking is a human concern transferred to animals without any basis whatsoever. It does not exist in animals who are absolutely incapable of evolutionary reasoning. Competition between different members of the litter starts early, and hierarchy is quickly established without maternal interference. While the mother is taking care of her own biological requirements, the litter, or at least some of those that can satisfy those maternal requirements, also will survive.

Olfactory attraction is important in animal mother-infant bonding (Nadler, 1974). Sniffing and licking the neonate is continuous. The attraction of the animal neonate to maternal mammary glands is also olfactory. Animal mother and offspring attract each other in some ways that appear difficult to explain, at least by human standards. Sniffing between mother and offspring contributes to the closeness of the animal family unit. It is only at puberty that the olfactory identity of the offspring is altered, not recognized by the mother as a part of herself, and therefore the offspring is rejected (LeBoeuf et al, 1972). This olfactory identity may have multiple implications. For instance, when an offspring has lost its mother, it will be adopted more readily by an "aunt" or a "cousin" than by an alien mother.

Taste, like scent, also plays an important role in an animal's mother-infant bonding. Immediately after birth, the mother continuously licks the newborn until a "taste" identification is established (Jolly, 1972; Rosenblatt and Siegel, 1981). After birth, infant licking continues to play a role in identification. The important point to be made here is that when an animal mother continuously licks and sniffs her offspring, an observer would immediately think she is doing it out of maternal love. The maternal love is certainly there, but licking and sniffing are only modes of an animal relationship, the way a human mother would look at, talk to, or touch her infant.

Touch also plays a role in mother-offspring bonding. Without getting into its biological or psychological meaning, a common observation is that animals like, or need, to touch or be touched as often as possible. The animal female, in her role as mother, is in contact with her young ones and the need to be touched is more than satisfied as she has an endless source of physical contact. The little ones nurse from the mother, climb on the mother, and touch her continuously in many different ways. It is highly possible that continuous touching of the mother's body by the offspring may have a biological function similar to the stimulation resulting from the offspring suckling. General skin stimulation may contribute either to persistence or prolactin secretion and estrus inhibition. Since animals like to be touched, motherhood offers a unique opportunity to gratify this desire. Animal infants, unlike adults, usually do not resist physical contact, a reason why others, beside the mother, also attempt to touch them. Numerous books (Ackerman, 1990; Montagu, 1971) have been written about the effects of touching.

Gregarious instinct and hierarchy also play a role in the animal's mother-offspring bonding. It is obvious that the presence of offspring gives the mother the double advantage of satisfying her gregarious instinct and establishing her dominance in the hierarchy of the family unit. This may be one reason why many animal females, once they have a litter, tend to isolate themselves until the end of motherhood. It may be the reason for the persistence of some mother-infant bonding even when nursing has been discontinued.

These five factors (lactation, touch, scent, taste, gregarious instinct) are essentially the basic biological elements of mother-infant bonding in animals and they are activated during animal maternal care at birth. This list is by no means complete and the existence of other biological bonds is not denied. These biological factors are intense, imperative, and almost exclusive at birth, but the bonding becomes more complex and goes beyond the biological needs as motherhood progresses.

When these biological factors are altered, the bonding is in danger. For instance, separating the mother from the offspring for only a few days is all it takes to disturb the animal bonding (Abitbol and Inglis, 1995). The bonding may persist anyway, but it is certainly diminished or at least altered.

Mother-Infant Bonding in Animals; A Touchy Subject

In terms of mother-infant bonding where the animal is considered by some as by far superior to man, the love, care, devotion, and attachment of the animal mother for her offspring has been idealized and regarded as a state of perfection which humans cannot match. Whenever I discuss the subject of mother-infant relationships in different animal species with scientists and nonscientists, my interlocutor often becomes emotional and even angry in defending the purity of this relationship, especially if the interlocutor owns a pet that recently gave birth to a litter. I have asked many psychiatrists why discussion of this subject brings such an emotional response on the part of my interlocutors. The psychiatric explanation always has been the same: I have activated deep-seated unconscious conflicts that are difficult and painful to bring up. Like any other human, the interlocutor is involved in frustration, anger, and even guilt, all deeply buried in the unconsciousness. The main manifestation of a strong unconscious activity (the mother-infant relationship being the strongest in each human) is through projection unto others; the animal mother-infant relationship, because of its simplicity, is an excellent receptacle for any projection, transfer, or displacement.

Through the years, I have learned to be very careful whenever I allude to my reservations regarding the pure love manifested by the animal mother toward her offspring. I do it progressively, and I stop when the conversation becomes too tense. I slowly introduce my comments on the matter of mother-infant bonding as reported in this chapter. I always start with my observation that the maternal behavior of the animal mother toward her offspring is very simple in its manifestation, always stereotyped, with no variation whatsoever. Licking the neonate at birth is stimulated by scent in addition to love on the part of the animal, at least in the early stages. I stress the biological aspect of breast feeding: the only way the animal mother can empty her painfully distended mammary glands is through suckling by neonates. Of course feeding her progeny is also a maternal concern. The distress the animal mother feels when she has lost her litter is due to a disruption in the mother-infant bonding, but also to painful mammary gland congestion since artificially relieving this congestion provides improvement in the animal distress. The animal mother never helps the litter get near her nipples, does not individualize care for different members of the litter, and finally, cannibalism of neonates by the mother, the father, or other members of the same species is not unusual. All these facts are well known in the medical and even general literature.

Concept of Love in Humans

I will first attempt to define emotion and love in humans to determine how much our conception of them exists in animals. At the elementary level, human love is the ability to perceive and participate in the emotions of another individual. The feeling of one individual also becomes the feeling of another individual and the emotional boundaries of an individual extend to and involves another individual. Therefore, there is some kind of communion, a sort of emotional fusion. A person is capable of love when he or she can feel another person's happiness or distress as his or her own. There is no love when one is limited to only one's own feelings, unable to appreciate the feelings and emotions of another; when one becomes blind and unreceptive to another's feelings; or when one manages to separate oneself from another's emotional situation. This simple definition of love has an important dynamic consequence: one contributes to his own happiness or sadness by contributing to another's, and one helps another in distress because one feels another's distress as one's own. It almost appears that one is relieving his own sadness or distress.

An important aspect of love, as perceived by humans, is that it is unconditional; it is not contingent on any outside circumstances. A mother loves her child regardless of whether the child is healthy or not, whether it fulfills expectations or not. When a woman loves a man, whether he is successful or not, faithful or not, it can make her happy or unhappy but it will not change her love for him. At least it take times for human love to change.

Another important criterion of human love is that, for each case of love between two individuals, a very personal character makes it different from any other case of love. A human mother, for instance, loves all her children intensely, but for each child a personal and specific character makes it different from her love for any other child. In other words, human love is never stereotyped, but rather individualized for each specific situation.

The last important criterion of love is its permanence. A love relationship may terminate physically and the two lovers may no longer see each other but it leaves a permanent mark or a scar in the memory of the former lovers.

Love at the human level is certainly much more complex than the simple approach I am proposing here (love = awareness of another's feelings; love is unconditional, individualized, and permanent). This rather philosophical approach has been drawn from Martin Buber (1970) and Sigmund Freud (1963). I adopt it, not for its completeness, but because it is sufficient to approach animal observation since we are limited in our mode of communication with them.

Emotional Mother-Infant Bonding in Animals

The human mother also experiences biological bonding as she breastfeeds her infant, touches it continuously, enjoys its company, and in extreme cases, sniffs it and even licks it. However, the presence or absence of this biological bonding, does not measure the intensity of the human bonding, the way it would in animals. A human mother may not breastfeed, may return to estrus, become pregnant again, be separated from her infant for many days or years, or the offspring may reach puberty, yet the mother-infant bonding remains practically unaltered. It is much more than biological.

Love in the mother-infant bonding of animals is obvious mostly in terms of biological maternal behavior (nourishing, sniffing, licking, touching, etc.). Many of our present concepts of human love and mother-infant bonding are not so obvious in animals. However, if the previously outlined specific criteria of human love are used, it is doubtful that they exist in animals.

Among mammals, offspring rejection is a common practice

and is the rule in wild life (Klopfer, 1971). This is done in various forms, depending on the species, and at any stage during the growth process of the young ones. The reasons for this brutal rejection are obscure and not always understood, i.e., when the litter is too large or the young ones too gluttonous and the mother does not produce enough milk. The demand becomes greater than the supply so the demand has to be reduced. It also is possible that suckling an empty mammary gland can be painful, leading the mother to take drastic measures.

The return of estrus is usually associated with litter rejection because the hormonal pattern of lactation and estrus are opposite. If for some reason the pituitary hormones start to activate the ovary for estrogen production, not only does estrogen inhibit prolactin, thus putting an end to lactation, but the female also goes into estrus (Pond, 1977). An animal female in estrus is anything but motherly (Schaller, 1972).

When the offspring start looking for food to supplement breast feeding, different behaviors, depending on the species, become apparent. In general, the offspring is well accepted and tolerated by the mother who does not compete with the young ones in their search for food. However, there are many exceptions.

One reason for rejection is that when the food supply is low, the young ones are the first ones to die or be killed. Competition for food between mother and offspring is common among animals. Typically, the mother lion will barely let her litter approach a kill until she is satiated first (Gubernick, 1981). An animal mother never deprives herself of food to feed her hungry offspring. Competing for food against the voracious appetite of growing youngsters can be a rejection factor. A hungry animal mother is no longer motherly. Also, starvation inhibits lactation. Safety also can play a role: for instance, constant attacks by predators make the presence of offspring a liability when one has to run fast. Finally, offspring are frequently rejected for no reason whatsoever. There are exceptions: pregnant animals who are also lactating and nursing offspring. These exceptions are rare, not always proved, and mostly seen in animals living under human supervision (Wilson, 1975).

When a young animal leaves, it is not voluntary but, rather, it is rejected by the mother. The rejection of the offspring who have barely reached puberty, is brutal (Ewer, 1968). Eventually, the mother-offspring relationship has to be discontinued, and this can occur any time after birth and before adolescence. This is accomplished in many ways: the cheetah will not return to her litter; the sow will eat some of the piglets; the mare will bite or kick the pony to death; the ewe will not allow suckling, etc.

Although biological attachments to offspring are preponderant, but emotional attachments in animals are not denied. I have personally observed the following example of emotional attachments in animals.

A female dog lives with a loving human family where it is apparently happy in view of the care and the constant petting she gets, and the way she wags her tail to welcome every member of the family. She has a noticeably distinctive relationship with each member of the family. One day she becomes pregnant, gives birth to a litter, and raises them the way a bitch does. During this period she is withdrawn from the human family and is very protective of her litter, barely allowing anyone to touch them. However, as the puppies grow, they no longer suckle and they feed themselves independently. When they are given away, the bitch has no objection whatsoever. Once the litter is gone, the relationship the female dog had with the human family returns to what it was previously. Later on, if the bitch is exposed to a member of her previous litter, she behaves indifferently, the way she would behave with any dog on the street.

Analysis of this situation is very simple. The bitch developed an emotional relationship with the human family. There was a temporary cooling off period because of the biological upheaval of pregnancy and motherhood, but once these events were over, the emotional relationship resumed. The relationship with the litter was mostly biologically dependent (lactation, scent, etc.), and as soon as lactation ceased, the biological relation with the litter was over. The bitch was capable of biological and emotional attachment. Her attachment to the litter, although intense, was biological and temporary, while her attachment to the human family was emotional and permanent. Of course, this situation has been simplified for better understanding. The truth is that the bitch was attached to the human family for biological reasons, too, since they fed her very well. The bitch also can be emotionally attached to her litter.

The lack of emotion expressed by an animal mother when the offspring is in danger can be startling: a wild goat climbing uphill will not slow down or return to her kid who is stuck between rocks (Gubernick, 1981). A female kangaroo will not help her tiny offspring creep and climb on her body (Ewer, 1968). A sow will not help a weak piglet who has been displaced by stronger piglets reach an udder. A mother chimpanzee has been observed to nonchalantly watch her own offspring taken away from her and eaten by other chimps. In practically each and every species (Goodall, 1971), one can see instances where the animal displays this lack of emotion regarding the offspring.

Among carnivores, the relationship between the litter and the father (assuming it is known) is too often idealized. The mother's usual concern is to prevent part of the litter from being eaten by the father.

Life in the jungle is not always the paradise some people think it is and dire situations constantly occur that are not exceptional but daily occurrences. One hears more about a monkey who steals an offspring from a mother to give it maternal care, a rare and even doubtful event, than about the constant stealing of offspring between monkeys of different species, or even within species, for the purpose of eating them. Cannibalism of the young by the mother or any other group member is often observed among different mammals even in captivity (Ewer, 1968). This practice must be even more frequent in the wild since there is no animal keeper to prevent it, and they are more exposed to starvation.

Differences Between Animals and Humans in Mother-Infant Bonding

The love of the animal mother for her offspring is not always individualized because it is essentially stereotyped like everything else in animal life. The lioness, for instance, will deal with different individuals of her litter and also with successive litters in an identical way. This is true for lionesses all over the world, and for all past and future generations of lionesses. In animals, this relationship is essentially biological, although elementary emotions can be encountered, as previously reported. In the human mother it is biological too, but only minimally; it assumes an emotional and spiritual dimension far beyond the capability of animals.

From the offspring's point of view, the relationship is terminated as soon as it has learned to survive independently; from the maternal point of view it becomes less necessary when the biological determinant ceases. Therefore, the crucial difference between animals and humans is that in animals the relation weakens when biological needs disappear, while in humans it persists beyond biological needs.

Although animal species may be caring (beyond biological needs), it is always rudimentary and brief; there is no comparison to the exceptional and various amount of care a human infant needs from its mother for such a long period of time. It is much greater not only in duration and quantity but also in quality. No animal offspring born with the same degree of prolonged helplessness as the human offspring will ever survive if taken care of by an animal mother. The animal offspring learns to search for food and avoid danger more by copying the mother's successful actions rather than from maternal active teaching, as is commonly thought. On the human side, maternal teaching continues usually until, and even beyond, adulthood. At least it has the potential to be that way. To the contrary, the young animal steps into the world under the Darwinian law of the jungle (survival of the fittest, the strongest, the fastest) too early. As stated by Gould, "nature has no morals."

In the animal mother-offspring relationship, there certainly might be many examples of positive characteristics besides the crude biological facts as reported here and positive emotional expression. In humans also, biological relationships are present as previously reported. Therefore, the description presented here is by all means incomplete, and there is certainly much more to learn in the mother-child relationship in animals as well as in humans. However, one already can perceive a biological aspect symbolized by animals, and a human trend which is biological but also emotional and spiritual, at least potentially. The demarcations are not always clear, but the orientations are different.

The relationship between human mother and offspring is complex, difficult and often very frustrating unlike the same relationship in animals which is elementary, simple, stereotyped and uncomplicated. Not only is it much more difficult to raise a human infant from a purely biological point of view, but we also expect more from the human infant. The animal offspring is only expected to go through a stereotyped life and live an identical life to that of the parents. In some cases it is understandable that the animal relationship may become attractive to humans.

Confusion always results from arbitrary transference of a human condition to animals. For instance, weaning in humans is completely different from weaning in animals. In humans, it means that the mother, even though she still has milk, has decided to separate the child from her breast to introduce it to new foods and, if anything, the child will need even more care after being weaned. In animals the mother weans the young because she has no more milk left since her mammary glands are no longer producing milk, and she does not want to be bothered with the continuous and useless irritation of her nipple. In fact, this weaning may signal termination or at least weakening of her relationship with the offspring. One can see the tremendous difference: animal weaning follows the end of the mammary gland function and may signal the end of the mother-offspring relationship, while human weaning is a voluntary decision in spite of continuous mammary gland function and in no way signals the end of the mother-child relationship. Similar observations are daily occurrences in animals, at least in wildlife.

Biological Versus Emotional Bonding

We can distinguish two types of mother-offspring bonding. The biological bonding, as previously described, is preponderant in animals at the beginning of their motherhood. It persists as such and is essential through their entire motherhood. The emotional bonding also appears at an elementary level in animals later on during motherhood and may be of varying degrees depending on the species. It is not as obvious or essential as the first one. In humans, it is the opposite, or at least it is different. The biological bonding is there from the beginning, but it is neither as essential or obvious as in animals. Here the emotional bonding is obvious, intense, essential. When one compares mother-offspring bonding between animals and humans, it is essential not to compare pears and apples. From the biological point of view, animals are perfect and far "superior" to humans. From the emotional point of view, animals may also be "perfect" at their elementary level, but in no way should they be compared to humans who have reached an intellectual and cultural level by far "superior" to animals. The purpose here is not to denigrate animals, who are certainly perfect as animals. The purpose, rather, is to not denigrate humans who have gone far in their evolution and are still struggling with their purpose in life (Teilhard de Chardin, 1957). Animals deserve respect and dignity. However, to equate biological superiority to full superiority is equivalent to negation of many millions of years of human evolution. Animals and humans are each superior in their respective domain.

Conclusion

The facts I have presented here may be hard and crude, and I apologize for shaking up some common beliefs. All students of animal behavior, including myself, have encountered numerous examples of tender loving care, maternal sacrifice, and absolute devotion that would appear contrary to what has been reported here. Most of these examples are certainly true, and I accept as a criticism that I stressed the biological aspect of the animal mother-infant bonding to better describe it and because it is often forgotten.

However, it would be an error for humans to turn to animals for guidance in terms of mother-offspring relationship and maternal love. Such a consideration would appear naive, but unfortunately it is often made, although mostly in nonscientific circles. Biological factors are not strong enough to determine the mother-offspring relationship on the human side. Humans need a spiritual and emotional dimension, a "third" dimension, which is encountered in animals only at an elementary level. Man should stop looking to animals, in general, and to the animal mother particularly, as some kind of divinity, inspiration, guidance or direction for our real nature, for our rejected origin that we should go back to, for a paradise lost with the hope to regain it. Animals are suitable only for animals' needs. If humans are to improve, and they need to improve at the mother-infant level as well as at many other levels, it will be through increased emotional relationships and not biological regression to animal level.

ANIMAL LIFE IN CONTACT WITH HUMANS

Introduction

Today, animals, specifically great mammals, live under two different conditions - in the wild (the jungle) or directly or indirectly in contact with humans. The first is decreasing, the second is increasing. Jungle life is vanishing as human population explodes and gradually takes over animal territory. The best example is the United States of America, which used to be the land of the buffalo, the grizzlies, the cougars, etc. This large animal population has been almost completely exterminated (Buffalo Bill claimed to have personally killed one million buffalos), and instead their land is presently occupied by 230 million human individuals. This leaves Americans with deep guilt feelings, and the country is now promoting legislation to protect whatever is left of this previously huge animal population. We idealize and fantasize about the beautiful life which we have deprived them of. We also export our guilt feelings abroad and ask the people of South America to save their rain forests for the monkeys; the people of India to save their jungle for the tiger and elephant; and the people of Africa to save their mountain forests for the gorilla. We want to prevent natives in foreign countries from doing to their animals what we did to ours. Tremendous efforts are being made today to save some territory (outside of the USA) designated as "national parks" for the purpose of preserving the natural way of life of these animals and saving the endangered species. The amount of animal life in contact with humans is gradually expanding as the animal territories are invaded. How human contact will modify animal behavior is a problem to be addressed, especially since certain animals (cattle, horses, and eventually higher primates) have almost disappeared from the wild and are existing only in contact with humans.

Now it is time to describe in detail the changes that occur in animal life when humans are present. Emphasis today is on the massive onslaught against multiple species (especially the great mammals and beautiful birds), the invasion of animal territory by agricultural expansion, the cruel life of animals in zoos and laboratories, and mainly the problem of endangered species. While serious attempts have been made to save some endangered species in their natural habitats by preserving their territories, unfortunately it seems that many of them will have a chance of surviving only in man-made or man-protected habitats. As the human population grows, the vast territories that belonged solely to animals are slowly shrinking, and eventually even land with arid climates will be taken over. The lion, tiger, leopard, elephant, rhinoceros, hippopotamus, zebra, giraffe, bear, etc., will be found mostly, or only, in cages, zoos, jungle habitats, or "national parks," under some human supervision. It is not so much that animal species in the wild are disappearing, but that they are being replaced by animals who live in contact with humans (cattle, pets, animals in zoos, jungle habitats, national parks, and laboratories). How much animal life changes from life in the wild to life in contact with humans at present and more so in the future, will now be discussed.

Changes Humans Bring to Animal Life

The first and most important change is food. An animal living in human contact is generally well fed. There is no such problem as waking up every morning wondering where the next meal is coming from, whom to kill, whom to steal food from, how many plants or how much grass is left to eat, or how many must share it. Starvation is unknown! Food is provided regularly; there may even be some variety to suit different tastes, and it is usually more healthy than jungle food because it is scientifically supervised.

An important human innovation introduced in animal life is regular mealtimes. There is no such regularity in jungle life. Wild carnivores know only two conditions: being hungry and hunting for food, or being satiated and not caring about food. There is no in-between stage and the duration of the rest period between meals depends on how abundant the last meal was. The meat eaten by carnivores is so rich in calories there is no need to eat continuously. In a human environment, however, the animal does not go hungry and can, therefore, sleep most of the time. Conversely, wild herbivores eat food so low in calories they have to graze almost continuously. Herbivores in a human environment are given food high in calories (grains) and they also can go through extensive periods of inactivity. The result is that animals who live on farms, in jungle habitats, zoos, and so on, are inactive most of the time. This may become a vicious cycle, since inactivity leads to less need for food and therefore more rest. These animals appear to be continuously bored. I have no quarrel with this concept except perhaps this is boredom by human standards; we have no real proof the animals are not enjoying this relaxation. Solving the food problem for animals is an innovation brought by human contact.

Intra-killing (within a species) or inter-killing (between species) also has vanished because competition for food is no longer necessary. Besides, intra- and inter-killing is not allowed in human contact. The skill for killing is quickly lost when animals are captured in adulthood and is never learned when they are born in captivity. The return of captive animals to wild life is associated with serious problems, especially for those born in captivity.

Human contact eliminates the need for animal population control. Cattle or poultry are slaughtered and used for human consumption, so that the demand is more or less equal to the supply. For other domestic animals, techniques such as castration and birth control, study of the fertility periods and artificial insemination are widely used. Under no circumstances does this animal population get out of control. In fact, it may be reversed because many species have difficulty reproducing when in human contact. Veterinarians, zoologists, and geneticists have made remarkable progress in the study of animal fertility, although many questions are still unsolved. It is well known that in zoos or animal parks animal males do not frequently look for coitus. A common error humans make is to think that all animals in captivity should be able to mate and reproduce, forgetting that in the jungle it is the privilege of an elite minority. With so many species living in close proximity, it is also possible their pheromones are altered or mixed up and individuals become immune and less responsive to their stimulation. In wildlife, sex is achieved through hierarchy, dominance, female submission and abuse. This hyperactivity may be responsible for excessive adrenaline secretion and production of other hormones, which play a role in animal sexuality. Animals in contact with humans have a different lifestyle, which is essentially quieter, and this may be responsible for less hormone production and less sexual activity.

The life span for animals in human contact is dramatically extended (except when they are slaughtered) and, unlike life in the wild, old age for these animals could become a reality. However, there are exceptions due to crowded conditions, limited space, or life in cages, and some animals, especially those captured in adulthood, have difficulty in adapting to a new life and do not survive.

Hierarchy and dominance may still play a role when animals live together in zoos or parks, but these become less apparent because competition has disappeared. Anyway, as soon as animals start to fight for any reason, they are immediately separated.

The mother-offspring relationship is changed, and protection of the offspring is not acutely necessary. Competition and search for food is eliminated, and since the mother is well fed, lactation may last longer. With humans continuously patting and playing with mother and offspring the relationship is more peaceful and long-lasting. As soon as any aggression appears, mother and offspring are separated. The transition between childhood and adulthood is easier than the brutal transition in wildlife and is one of the greatest advantages of the human presence. This is best demonstrated with the cat, dog, horse, etc.

Adolescent animal care is now common. For instance, a cat or dog litter will get some adolescent care not only from the mother in a limited way, but also from a human family member who strongly supports the mother and conditions her to extended care. The separation is peaceful. The Darwinian law of natural selection and survival of the fittest applies much less.

The greatest shortcoming for animals in contact with humans is the problem of territory. Animals are accustomed to open spaces where they run freely, and now in some extreme cases they are confined to cages with barely any space to move. This territory limitation is indeed cruel if animals have to stay in cages for life. However, this situation is slowly improving and animals can be seen moving freely from cages to more open spaces. A better solution and one being implemented today is the creation of natural parks and jungle habitats where the territory requirement is partly solved, while the advantages of human presence are kept. With food provided abundantly, the vast open space necessary for hunting purposes would no longer be as necessary.

Human cruelty to animals is presently a subject of continuous discussion and controversy, and various organizations, such as the American Society for the Prevention of Cruelty to Animals, monitor this. I am not excusing or ignoring it, but it is important to stress that the cruelty is judged by human standards. If a human being could imagine living the life of an animal in the wild under Darwinian conditions previously described (onslaught of population, fierce competition for food and territory, hierarchy based on strength, "kill or get killed"), then, by human standards, this would be real cruelty. Some animals, especially females and the lowest ones in the hierarchy, may experience less cruelty with human contact. For some animal individuals, the cruelty of jungle life easily can match and even exceed human cruelty.

Today, standard techniques are used by humans in approaching animals, at least under ideal situations. It is important to reach the animal as early in life as possible. This does not mean that an animal has to be separated from the mother, even though excellent results are obtained in the case of a mother's death when one has the facility of an animal nursery, and when bottle feeding is available. Constant touching and playing are crucial, since touch and gregarious instincts are so important to animals. Carefully touching, patting, and caressing animals is an art, and this science of animal-human relationship is called conditioned behavior, as described by Pavlov. It is unbelievable what animals can achieve if they feel love, care, and concern: animals are very intuitive and perceptive in that matter. Once the basic needs of life are fully satisfied, there is the possibility of giving up aggression, at least partly. It is because the total attention of the animal is not concentrated on the basic needs that their behavior changes. Veterinary medicine, animal psychology, and people devoting themselves to improving living conditions of animals have contributed to this progress.

Animals have a problem trying to understand humans. In a way, they are puzzled, fascinated and frightened, but they are also attracted to us. No mammalian animal ever unconditionally gives food to another animal, except for the mother to her offspring, and this may be only a biological drive. Who are these strange creatures who under ideal conditions freely give you everything you need: food, shelter, attentive and skilled care when weak and sick, companionship, and all the basic needs for life? Why do they do all this without struggle, killing, or asking anything in exchange?

Some animals get used to human companionship and can even close their eyes to such things as limitation of territory and lack of freedom. When an adult animal from the jungle is caught by humans, its adaptation to life in contact with humans is difficult, if not impossible. But when an animal is raised from birth in human contact, it is almost impossible for that animal to return to the wild. (see the films "Born Free," and "Story of the Grizzly"). Animals raised by humans never freely return to the wilderness and when forced to do so, they usually try to come back. When an animal becomes attached to a human being, this attachment can reach unbelievable dimensions on the part of the animal for humans, especially for the specific one who has skillfully created the relationship. It is in this unique relationship that the animal displays devotion, loyalty, sacrifice, and, let us not hesitate to say it, love. This type of relationship never exists between animals, and even in the mother-offspring relationship it is debatable because it is temporary. It is wrong to think that animals are capable of this kind of relationship on their own.

When discussing the relationship between animals and humans, the cruelty of humans versus animals always comes to mind. This is not denied but the opposite is also true. I would like to report some isolated observations made by me as well as others.

Shannon Brownlee (Discover, Science Dec. 86, p. 70) reported on a bottlenosed dolphin living in a small enclosure behind bars in Florida's Sarasota Bay. A number of bars on the enclosure had rusted away. The dolphin not only never left the enclosure through the wide opening, but once a manatee that swam into the enclosure was promptly chased out by the dolphin.

Mary McLaughlin (New York Post, 8-8-1991) reported on a couple of Japanese snow macaques (a male and a female who was pregnant) who lived in a cage in a home located in Long Island, New York. One late afternoon they escaped from the cage, but the following morning they returned to their owner's arms. Years ago, I owned a canary who lived in a rather small cage. One day it escaped, but to my amazement it came back a few hours later and on its own reentered the cage through the door which I had left open.

Our cat "Minou" was a fantastic animal living half indoors and half outdoors. It would follow us whenever my wife and I took a walk around our house. One day we walked a few blocks away from our home to visit a neighbor. A few hours later, while still visiting, some person mentioned that there was a cat standing outside the entrance door. It was "Minou" who had followed us and had been waiting for us all that time. When we came out, he welcomed us with "angry" meowing for being so late.

Similar cases have been reported by numerous observers. I sometimes wonder if animals living in the wilderness are really free, or whether freedom, as conceived by humans, really even exists in wild animals. Animal hierarchy is nothing more than natural slavery. Even the individual at the top will not be there for too long, because competition is fierce and he eventually will be displaced.

While we should make every effort to preserve and maintain animal wildlife, more and more animals live in human contact, and it is doubtful this can be changed. I sometimes wonder if we have not created new animal species, completely different from the ones living in the jungle. A cat is different from a lynx, a dog from a wolf, a cow from a buffalo, etc. This list is expanding and may include in time to come all the big cats, bears, maybe all apes and even monkeys. A present policy for preserving endangered species of South American monkeys is first to raise them from birth in artificial conditions, then try to introduce them to jungle life. The assumption is that they would prefer the freedom of jungle life. I have no quarrel with such an assumption, except that it has yet to be proved that a permanent life in an artificial human habitat is not better. I have only a vague idea about how the Neandertals lived, and I will not argue with whoever wants to return to their lifestyle, provided this return is not imposed on me or others.

Ideal Animal Behavior in Human Contact

What is going to be described is almost wishful thinking; it already has been observed in higher primates (chimpanzee and gorilla) and in some aquatic mammals (dolphins). I will attempt to describe how some animals will react to human contact under ideal conditions. I say ideal conditions because it happens only when extreme efforts are made by human trainers in eliciting a new type of animal behavior completely different from the one in the jungle. Animals will show tremendous changes indeed. Because an animal has established an ideal contact with humans, most if not all the conditions that made its previous way of life mandatory disappear, and aggression and competition become less and less a way of life. How is the animal going to behave without fierceness and deadly competition? The scientific and the general literature often report on isolated cases of animals showing intelligence and emotional maturity. I repeat that it is only under ideal human conditions that animal intelligence and emotion really may appear.

Through conditioned reflexes, man encourages any sign of intelligence and emotion in animals by rewards. Intelligence and emotion are greatly appreciated by humans who are very eager to see these values in animals. These values which are practically meaningless in wildlife suddenly take on a new dimension: a dolphin who is intelligent, a cat who is smart, a chimp who learns to communicate, a dog who is loyal and a horse who is faithful are animals all displaying unnatural qualities. These potential qualities are shown only in the presence of a highly skilled specific human individual who, through his/her intelligence, sensitivity, patience and understanding has allowed this potential quality to develop in the animal. Even so, these qualities are only skin deep because one does not erase a billion years of animal behavior by spending months or even years with an animal. What it took millions of years for man to acquire can be assimilated only superficially in a short time by animals. Thrown back in the wild, the animal returns either to its inherited behavior, for which it has all skills and predispositions, or it dies. The most one can expect in the average animal, even after prolonged human contact, is a mixture of a large dose of jungle instinct and a minimal dose of humanization.

To think that human contact has produced a revolution in animal psychology, a disappearance of Darwinian aggression, and replacement by emotional values is pure fantasy. The man who thinks he completely controls a dog may one day find himself bitten by that dog, and the man who thinks he has made a friend with a bear may one day be torn to pieces by that bear. Animals are always animals, and humans do not affect them deeply. At most we create a conflict: a conflict between their animal nature and the human encounter. When a dog who lives with a human family meets a pack of wild dogs "inviting" him to join their jungle life, he is facing a conflict.

No matter how critical one is of human life, there is no doubt that, for better or worse, each one of us tries to give a unique feature to their life that is different from anyone else's. This is a fundamental difference from the stereotyped animal life. Animals living in contact with humans receive some elementary personality: two wild dogs struggling for survival are essentially alike in their savagery, while two pet dogs living in different homes acquire some elementary and different personalities. Two chimpanzees living in the jungle are different only in terms of their hierarchy, but two chimps with whom two different psychologists are trying to communicate with have different personalities. Human contact creates some kind of personality in animals. Present living conditions for some animals are creating new situations that never existed previously and now need to be addressed. Goodall (1971) and Fossey (1983) have described some elementary "personalities" among their subjects, but I wonder how much their presence influenced their subjects.

Animal Potential in Human Contact

Perpetuation of animal species is the third most important international problem today. (The first one is the prevention of nuclear holocaust, and the second one is human death by starvation.) The question I would like to ask is: from an animal, not a human, point of view, which is the better life for an animal, the wildlife of the jungle, or man-made life in zoos, national parks, and artificial jungle habitats? The major consensus so far is that jungle life is better than life in a cage, zoo, circus, or a laboratory. However, if animal life in natural parks, and so on, could be improved so as to eliminate the feeling of being caged, then the answer should be reconsidered. As stated by Diamond (1992), the future of the animal species is, one way or another, in contact with humans. The concept that animals are born free could be a misconception since not one single aspect of wild animal life could be considered free by human standards. Animals living in contact with humans also have "rights," which may be different from those of animals living in the jungle: to assume that they desperately want to return to the jungle or that their lifestyle is degraded compared with jungle life needs to be documented. As previously mentioned, humans who are frustrated with their lives have a tendency to fantasize about the freedom, pure sex, and deep mother-offspring love that supposedly exist in wild animals. The facts presented so far demonstrate beyond a doubt that this is not so. That wild animals have a better life than animals living in contact with humans (in ideal conditions which are not always so) needs to be documented. There are shortcomings as well as advantages and disadvantages in both. Realistically, we must accept the fact that wildlife is shrinking in spite of all efforts. The only way to preserve some endangered species may be under improved living conditions with humans. As Diane Fossey (1983) understood only too well, the allocation of fertile land to wild animals is becoming more and more difficult in the face of the growing human population. In India, each individual tiger in the jungle requires fifty square miles of territory; when properly cultivated such an area could feed a whole town. To permanently save the great mammals, it may be cheaper and more practical to look for another solution.

Purpose of this Discussion

Why such an in-depth discussion of such a sensitive subject? The main purpose of how animals can assimilate human behavior is an attempt to understand how humans, originally living in an animal world, have struggled and partially succeeded in reaching human values. About three million years ago, man started a long journey on a tortuous road beginning with a condition similar to today's ape, and finishing with the present *Homo sapiens*. To describe the chronological step-by-step intellectual and emotional transition from ape to man is a tremendous undertaking. However, it can be stipulated that, somewhere along the line, starting with the australopithecines or with *Homo erectus*, aggression and hierarchy, which are typical of animal society, started to ease with the possibility of food storage and cooperation between individuals. Using this hypothesis, it can be speculated that a glimpse of what the ape-man behavior was could somehow be revived by observing the present apes or any other mammals, not in their natural jungle life but in the way they presently react to human contact. To those researchers who are trying hard to communicate with a chimpanzee or a dolphin through some sign language, I offer the following: jungle animals do not communicate because there is no need to. Hierarchy and aggression, their only modes of relationship, are sufficient. Only with humans do animals feel the need to communicate. After all, the first glimpse of humanity began when trust started to take the place of distrust, cooperation the place of aggression, communication the place of hierarchy, language the place of screaming. Is this not what we observe in these communicating primates, skilled dogs, and sensitive dolphins? Are not these animals also starting, through human contact, on the long road of human evolution, by looking at the dim light of humanity? The australopithecines after all had no more of a brain than an ape or a marine mammal. My educated guess is that the closest we could ever get to seeing the australopithecine way of life and even of early specimens of the genus *Homo* is simply by extensively training and observing chimpanzees and other primates in human contact. I propose an extensive study of the higher primates under improved conditions to recreate the way of life of the australopithecines and other extinct hominids.

I will not challenge the fact that my concept of the human role in the present animal condition is somehow idealized and influenced by wishful thinking, but so are our concepts of the present animal and human roles in animal life. For the last billion years, animals have been invading each other's territory and killing each other, and animal species have been appearing and disappearing. With or without human presence, endangered species always have existed (Moore, 1964; Owen, 1975). The present elephant has been preceded by numerous forms of "elephant" species that all disappeared without human intervention (Wexo, 1980). It is my usual joke that if *Homo sapiens* had been present at the time of the dinosaurs, he would have been made responsible for their disappearance.

All species are evolving, and we should question our motivations in our present attempt to freeze in time the living animal species, almost at the expense of the natural course of evolution. The present concept is that humans have a negative influence on animal life. I believe so, too. I believe also that invasion of animal territory without provision for placement of the "displaced" persons is pure genocide. On the other side, human starvation is rampant in many parts of the world and, with explosive populations, it has recently increased. Our planet today is feeding ten times more people than a century ago (Aries, 1982). It would be almost impossible to reserve extensive parts of the earth exclusively for animals and keep them off limits for humans. The evolution of the earth today is such that all its parts are interconnected (Teilhard de Chardin, 1964), and I do not see how it is possible for some parts to be put aside. Somehow, sooner or later, the future of the great mammals has to be related to humans. Here, man could have the heroic role of really saving some species which would disappear anyway. I do not see anything wrong in the survival of some animal species in human contact and under human supervision if there is no other way, even if their way of life is altered. It may be fantasy on my part, but the solution seems that animals have to become "economical" with their territories, even when they need help with food and have to give up, at least partially, the wide open spaces.

THE CRYSTAL BALL

Introduction

When Charles Darwin wrote "On the Origin of the Species" in 1859, he dared not include that man "descended" from the apes. However, it was so obvious from his work that this conclusion was made by others for him, and he became the most ridiculed scientist in the history of science. I suppose that Darwin himself could not figure out the future role of *Homo sapiens* in the history of the universe. Others did it afterward, among them Huxley (1946) in "Brave New World" and mainly Teilhard de Chardin (1957) in "The Human Phenomenon." The powerful, creative, almost divine role of man in the future of evolution was already perceived by famous writers such as de Saint-Exupery (1979) in "The Wisdom of the Sand," famous philosophers such as Kierkegaard (1944) in "The Concept of Dread," and Buber (1970) in "I and Thou." The future evolution of the human species has barely been attempted by scientists because they have always considered that we do not have enough data for such an undertaking, and, at least for the time being, it would be pure science fiction.

Publications such as "Man: 500,000 from now" by Robert Shapiro in 1933 and the "Future of Man" by Teilhard de Chardin in 1964 are few and far between. The latter is unusual in that it approached the problem by completely refusing to consider that the present *Homo sapiens* is permanent and final, and that human evolution will resume its natural course. Man is here to stay.

I have previously demonstrated that the human approach to the future of animal species is not necessarily pessimistic and that an animal-human "alliance" is possible. The issue to be addressed in this chapter is the following. Given the more and more extensive role that man is playing in the universe, what is the future of animal species and how is their "natural" evolution going to be altered? To put it bluntly, what kind of new animal species are we going to have in the future, in view of human interference, assuming any will survive. I consider this problem to be acute because it is not for the distant future, but for the present. This crisis is associated with tension, anxiety, clashes, and extreme views. It is a painful subject to face, but very soon it will be more painful not to face it. My attempt to answer these questions will be based on historical events.

Historical Considerations

Since the appearance of life on earth about half a billion years ago, the land, skies, and oceans always have belonged to vegetation and animals. Darwinian laws (natural selection, survival of the fittest) regulated life on planet Earth.

During prehistory, which extended from the early appearance of prehumans (2-3 million years ago), to the Neandertal (50,000-100,000 years ago), the part of the planet humans occupied was insignificant, although starting with *Homo erectus*, they scattered in small tribes to every corner of the Old World.

Starting with human history (approximately 10,000-20,000 years ago), human cultures took form and occupied Europe, the countries around the Mediterranean Sea, Southeastern Asia, and many other areas. These lands were taken away from the jungle and the animals occupying them. In the rest of the world one could find scattered tribes living in "harmony," or rather in equilibrium with animals.

During the last centuries, with progress in agriculture, industry, and health, an explosion in human populations occurred. Hunting increased and species such as the buffalo and whale became endangered. Raising cattle and poultry has practically replaced hunting. More land was needed for human habitats and sophisticated agriculture. The human assault on forests, jungles, and open lands started and continues at an increasing speed. Practically all wild mammals and exotic birds became endangered and some have even disappeared.

Isolated individuals, then societies, became concerned about the deteriorating conditions of animal life. This movement started simply, like the American Society for the Prevention of Cruelty to Animals. Then they became active, like "Protection of Wild Life," and pushed forward legislation that tries to preserve whatever is left of the animal kingdom. Presently, extremists are appearing on the scene. When I say extremists, I do not mean to be judgmental; I am only describing certain attitudes that may or may not prove justifiable. The fur industry is threatened to extinction; people are not allowed to build on land occupied by wildlife or with trees; a homeowner is taken to court for killing a rat eating vegetables in his garden; mouse traps should not kill a mouse but only capture it and then release it. Brigitte Bardot, the French sex symbol, wrote to the Pope asking him to save the dogs in Sarajevo without mentioning any concern for "ethnic cleansing" or genocide. I personally know a homeowner who expressed concern that many raccoons were occupying his backyard and might harm his children. He was told by the local animal society that the backyard belongs to the raccoons as much as to him, that he should learn to live with them, and should he harm any raccoon he would be taken to court.

Saving the rain forests of South America, the high hills of Central Africa, or the jungles of South Asia for their respective animals is an undertaking performed mostly under the influence of financial and political pressure from the USA and for tourism. It is not so much a local concern in these foreign countries.

Evolution goes on, and one has to distinguish between the human role in this disappearance on one side (and this is not denied) and the natural evolution under which some of these species may be near the termination of their cycle.

The reasons humans want all the animal species to survive in their environment are multiple. There is a human consideration for a fellow living creature. They are necessary for our education and understanding of nature. Animal science or zoology represents an important aspect of human knowledge and scientific research needs them. Although difficult to explain, I personally find that, if planet Earth became totally human without any animal kingdom, it would be frightening, emotionally disturbing, and even difficult or impossible to live in.

I have not been able to decide about animal versus human rights. Is a country justified in expanding its human population at the expense of animal population by taking the land and exterminating these animals? This question is not so much legal, rather, it is economical and moral. I do not know the answer, but what I do know is that countries who have already taken all the land from the animals during their past history (Europe, USA) are the ones urging those countries who still have large amounts of animal land not to take it over. These Third World countries are more and more independent-minded, and eventually they will act according to their own interests. Vast territories such as the South Pole for the penguins, the North Pole for some wolves and white bears, and some deserts, will be left entirely to animals only if they are not useable by man. But, as technology progresses and natural resources are found under the surface of the earth, and as economical pressures increase, even these territories will shrink. The voices of societies for the preservation of wildlife will be more vociferous but economic reality may (or may not) prevail in the long run.

Future Aspects of Animal Survival

Under these conditions, one has to look somewhere else for long-term survival for animals, in addition to preserving the wild territories for as long as possible. As a parenthesis, let us immediately state that these wildlife territories always will be, one way or another, directly or indirectly, under the supervision of or at least in contact with humans.

One field where human intervention successfully replaced natural evolution is definitely in agriculture. Primitive societies were and some still are food gatherers. We still see remnants of this condition like picking wild strawberries and truffles. However, the vast majority of food originates today from agriculture. Agriculture is nothing but human selection of wild plants that have been slowly and progressively adapted to human needs during thousands of years. Agriculture is nothing more than human selection interfering with and replacing natural selection, and it has been very successful. The process of human selection is still going on, and has reached a new and explosive aspect with genetic manipulation. Many endangered species of plants are now preserved in botanical gardens, and it would be possible to expand this project. Claude Monet, the French painter, initiated such a project in the garden of Giverny (France). This agricultural process can become the framework on which to base future human-animal relations.

The conclusion is that animal survival must be related to human existence. Besides, to freeze any living species in space or time is an evolutionary mistake and a misconception of life itself. Animals and humans continuously evolve in time and move in space. The necessary human conditions for animal survival and the necessary animal adaptations are numerous.

Humans will have to accept the presence of animals and, since they are more intelligent and technically more advanced, they will determine the conditions for coexistence and take responsibility for it. The greatest human contribution will be to provide or at least supervise food supply, which is the main determinant of animal life.

Effects on the animal way of life and even on their physical appearance are far reaching and very important. The Darwinian laws of evolution, that is natural selection and survival of the fittest, will be markedly attenuated and even disappear. The "future" nature will not necessarily select for survival those individuals who are the most fit in searching for food. Territorial imperatives, hierarchy based on physical strength, and aggression will have to cease.

Sex life will be entirely modified. Reproduction, based on the physically strongest individual appropriating as many females for himself as he can handle and eliminating the rest of the males from sex will give room to more "democracy" and a "monogamy-like" animal society, or at least a society where sexual activity will be more equally shared. The fundamental consequence is that the physically strongest will not necessarily be the selected one for survival.

Old age, almost unknown in wild animals who die in full maturity either by starvation or as prey, will occur and expand progressively. In general, the weak, sick, young, and old will be protected.

Lo and behold, some new qualities, unknown in the wilderness, will start to appear in animal species, because these new qualities will be appreciated by human companions: love, loyalty, pacifism, gentleness, intelligence, etc. The question is whether these new qualities (or any other qualities) will materially affect chances of reproduction. They may or may not. At any rate, the reproduction and survival of these specific individuals will be encouraged. Advanced science will step in and interfere in feeding, medical care, social activity, genetic selection, etc.

In summary, wild and savage beasts, mostly regulated by aggression and hierarchy will be replaced by more peaceful and more social individuals, each one with a specific personality. These new species will be different from previous ones mostly in terms of behavior but this may eventually reflect on their anatomy since sharp teeth, powerful claws, musculature geared toward aggression or speed, will no longer be modes of selection.

Imagining these animals in the future is not entirely wishful thinking and fantasy because humans have already undergone this condition. They started to implement this program in animals, and are already modifying animal species away from natural selection toward a human selection. This is how, over centuries, wild dogs and wolves have become modern dogs, wild cats and lynx have become modern cats, the ancestors of horses have become modern horses, the ancestors of donkeys (maybe a parent of the zebra) have become modern donkeys. The list is a long and varied one depending on the continent. The list is not longer because this adaptation-domestication takes many generations to be accomplished and happens only if there is a specific use and function for these modified animals such as cattle, poultry and horses.

Many other animals have partially adapted, such as those displayed in zoos and circuses: elephants, bears, monkeys, apes, snakes, aquatic mammals, parakeets, etc. The reason the adaptation of these animals is not as advanced as in cats or dogs is because human contact is established at the adult stage rather than from birth. But I am sure that if these animals had been in contact with humans from birth for many successive generations, today we would have different species of animals well adapted to humans such as the dog, cat, or horse.

Since the plan is already in place, there is good reason to generalize it to more species, especially the endangered ones. Some animals such as the tiger, lion, bear, shark, and even elephant, may be barely adaptable to humans or too dangerous for adaptation.

For these animals, national reserve parks or seaquariums would be the best solution. They would live in semi-wilderness under close or distant supervision.

Some others would be perfectly adaptable to humans, and their companionship would be appreciated by humans. This would include all the primates and many birds (pigeons). Surprisingly, more species can be added to the list: a typical example is the rhinoceros who, in national parks, has been found to be peaceful and of predictable behavior (much more than the unpredictable elephant) and has reproduced so successfully under human supervision that it has been taken off the endangered list. Special areas in the Key Islands of Florida have been exclusively reserved for some primates, and under close supervision have become successful colonies.

Raising animals on farms is here to stay, as it becomes more and more selective and specialized. Geneticists have pushed progress even further as specific features in these animals are selected.

The field of pets certainly can be extended well beyond its present limitation to cats, dogs, parakeets and turtles. I have seen people keep monkeys, bears, snakes and crocodiles as pets. Of course, this transition has to be done with tact, humanity, and dignity, and one has to be aware of any danger and all abuses. Special societies (already in existence) will always be involved in this matter.

I conclude that the future of evolution (for animals as well as other domains) is human selection versus, or rather concomitant with, natural selection. I am not taking sides, I am just describing both modes of selection. Natural (or Darwinian) selection is found to be attractive because it is natural. This means absolutely nothing. Darwin (1859) describes it as clumsy, cruel, and murderous. Gould (1977) talks about the hecatomb of individuals sacrificed for the sake of selection. I do not see anything wrong, therefore, in humans stepping in and, without entirely replacing natural selection, at least playing a role in it, and carefully attempting to modify it. This does not mean that wild species should be thoroughly replaced by domesticated (I prefer the word humanized) species. I am saying that in each species, a humanized one should be preserved parallel to the attempt to save the corresponding wild one (Hoage and Deiss, 1966). Mistakes will be made but eventual advantages can result for the endangered species. Some people may claim that it is tampering with nature, and in a way they are right. But at the moment hominids stood up and started to look around about three million years ago, they already planned to interfere with nature. *Homo sapiens* is considered to be a biochemical error, as stated by Sigmund Freud. *Homo sapiens* is an aberration of nature. Nature has always attempted to recapture this "rebellious species" but *Homo sapiens* always counterattacked: wearing clothes, mastering fire, building shelters and towns, developing agriculture and cattle raising, vaccination against diseases, health care, splitting atoms, inventing computers, traveling in space, genetics, etc. I do not see anything wrong in humans interfering with the natural evolution of the animal species since, according to the laws of evolution, in the long run they are all doomed anyway. I am convinced that, in the long run, this is the only way to save some endangered species.

ANIMAL SEXUALITY

Introduction

I now feel that I have accumulated enough material to objectively approach animal sexuality. The description of animal sexuality is based on many references listed at the end of this manuscript. The comparison between animal and human sexuality barely has been explored. It is difficult to find any major research that compares the two sexualities, because it is assumed there is little difference between them and sexual knowledge is easily transferred from one to the other. Such investigations would be revealing: regressive patterns to an animal level could be observed in abnormal human sexuality, or knowledge about orientation of human sexual behavior could be acquired.

During the past few years, intensive studies on the sexuality of mammals (especially primates), either in the wild or in protected environments, have been pursued. These studies have been facilitated by modern technology (high speed cinematography, modern videocameras, ultrasound, infrared, etc.), allowing close observation of the most intimate details of animal life. In addition, today there is a better understanding of the different hormones regulating reproductive endocrinology and sexuality.

This comparison of animal and human sexuality is made easier by the tremendous similarity of sexual practices among the different species of the animal world. In view of the great variety in anatomy, physiology, and behavior observed in animals, marked differences in their sexual function, as in other functions, would be expected. Yet, this is not so. It is not implied that all animals are identical in their sexual behavior, but that there are more similarities than differences between them. The technique used by an insect, crustacean, rodent, carnivore, or an ape for the sexual act is almost identical, monotonously similar. By contrast, human sexuality is very different from that of animals. The differences in sexual practice between an insect and a chimpanzee are noteworthy, and yet they become minimal when compared with the tremendous difference between a chimpanzee and *Homo sapiens*, as we will see later. This investigation will concentrate on mammals, including nonhuman primates, who are remarkably more similar in their sexual behavior to other animals than to humans.

Role of Scent in Animal Sexuality

As previously shown, animals rely mostly on scent and taste in their general life and this is also reflected in their sexual behavior.

Anatomically speaking, the primary sexual characteristics are the testes and the ovaries; the secondary characteristics are the male and female genital organs; and the tertiary sexual characteristics give either a male or female external appearance. Tertiary sexual characteristics are not always obvious in mammals, and a close look at their genital area is often necessary to determine their sex. Even when present, external characteristics play no role in sexual recognition between animal individuals as they do in humans. If animals have a sense of esthetics and if animal males classify females from a sexual point of view, it certainly would be based on an olfactory aesthetic and not a visual one. This is easily confirmed by observing two animals, of any species, about to have sex. The male pays little attention to the female body; he goes directly to the genital area and starts to sniff and lick. It is difficult to know how much information is gained by sniffing this area, but it already has been demonstrated that the species, the individual, maturity, and presence or absence of estrus are immediately registered. After careful "computerized" analysis of these olfactory data, a determination of the sexual profile of the female is made and the male decides whether or not to go ahead with the mounting.

Estrus in Animals

Estrus, or heat, is a marked swelling of the female genital area with labial tumescence and congestion around the vaginal opening. The estrus is seasonal in nature, occurring usually in the fall when the animal is well fed, but in some species it can occur any time during the year. It does not appear that the male in any way sexually stimulates the female or plays any role in the appearance of estrus in the female. Not all factors regulating estrus are entirely known, but they are definitely hormonal in nature. Briefly, the pituitary gland releases the follicle-stimulating hormone (FSH), which acts on the ovary and initiates estrogen production. Estrogen is the hormone responsible for the animal going into estrus. This stimulation occurs without the animal willfully initiating it or participating in its mechanism.

Estrus produces two important biological conditions. The first is tumescence of the genital and perigenital areas (the "pink lady" noted by Goodall, 1971). It is naive to think that estrus, which is very painful as evidenced by the continuous crying and screaming of the animal, is a pleasurable sensation enjoyed by animal females. The second is the production of pheromones by special glands usually located in the perigenital and perianal areas. Pheromones are present in the urine and feces as well as in vaginal secretions. These pheromones are volatile, can be sniffed at a great distance by males of the same species, and attract and guide the male toward the female's genital area.

Repeated coitus eventually releases the luteinizing hormone from the pituitary gland which acts on the ovary and the result is ovulation, progesterone production, and eventual impregnation. The painful estrus is usually relieved only when the female has ovulated and is eventually impregnated. The progesterone production associated with ovulation counterbalances estrogen and, therefore, relieves the estrus. The animal females do not have sex because they are attracted to a male or want to procreate. They have no concept of the relationship between the sexual act and procreation; they just passively submit out of fear of male aggression.

The male's need for coitus is also hormonal in nature. Briefly, the pheromones produced by the female in estrus are sniffed by the male and stimulate his pituitary gland to produce gonadotrophins which act on the testes and lead to testosterone production. The testicles become significantly enlarged, sperm is abundantly and continuously produced and builds up in the vasa deferentia.

The prostate also becomes congested and distended. The sperm components build up in these secondary male organs, which become painfully distended as evidenced by continuous screaming and crying, and leads to erection. This congestion and distention of the male organs is just as painful as estrus is in females and is relieved only by ejaculation. The search for a female in estrus, usually located by scent of pheromones coming from her perigenital area, then begins.

Not everything is known about sexual stimulation in animals, but it is clear that the scent of pheromones produced by a female in estrus has two effects on the male: first it sexually stimulates, and second it attracts and directs the male toward her genital area. The general contention is that, providing the male is well fed, in good health, and the weather is favorable, his secondary sexual organs fill rapidly when he sniffs the pheromones of a female in estrus. The desire for coitus is much stronger in the male than in the female. Once a female has ovulated, her estrus is relieved for the entire season, while an entire male population can be continuously stimulated by the pheromones coming from a single female still in estrus. The result is that the sexual organs of the animal male are continuously refilled after each copulation, resulting in distention and pain and, therefore, the need for repeated ejaculation. There are, therefore, always more males than females in need of sex. The lion, for instance, may look for coitus several times a day, while the lioness may be receptive to it only until she has ovulated.

Once the mating season is over, males and females behave indifferently toward each other, as if they were of the same sex. During the nonbreeding season, individuals of the opposite sex ignore each other for sexual purposes. There is neither masculine nor feminine behavior except for the short time devoted to sex. For instance, a male wolf who has just finished mounting a female wolf, will not, for that reason, let her share a kill. The important point is that a female not in estrus will not be sexually aroused by the presence of a male, and a male not stimulated by pheromones and with empty seminal vesicles will not be sexually aroused by the presence of a female.

It is easy to determine whether or not an animal is looking forward to coitus and veterinarians and all animal keepers know well the signs and symptoms of a female in estrus and a male in search of mating.

Selection of the Sexual Partner

How sexual partners meet, how long they stay together, and how faithful they are to each other is not very well known. Fundamental observations already have shown that animals (tigers, bears, cats) who live isolated lives may have one single sexual act and never meet again. Most of the time, however, the higher mammals, especially the primates, attempt to form a rudimentary social nucleus. A single male, or sometimes more than one, but always the strongest, organizes a group of females around him for sexual intent. It resembles a harem with frequent additions or subtractions, and often there also is a delineated territory. It appears there is no particular relationship between specific individual male and female. When the male indulges sexually with a female, there is complete indifference on the part of the other females. When there is more than one male in a tribe of lions, for instance, the male is completely indifferent to other males who are sexually indulging with a female. Two males can fight over a female, but only in terms of who goes first; as soon as the male has finished his sexual act, he becomes completely indifferent to what happens between the female he just mated with and the male who follows him.

Sexual jealousy is the frustration and anger of a human individual when his or her preferred sexual partner is involved or thought to be involved sexually with another individual. Such feelings are nonexistent in the animal world. No animal female ever expresses jealousy. When the male is displaced as the head of his harem by a stronger male, this is met with complete indifference on the part of the females. A female will never come to the rescue of the departing sexual partner. The male lion, sea lion, or baboon may fight for his females only because he is protecting his number of females and his territory rather than any specific female. A male who is not in heat will not attempt sex with a female whether or not she is in heat. If not in heat, she will not allow it because of a tight vagina causing a painful and almost impossible penetration.

Sexual Approach

It is most important for the male when approaching the female to eliminate all other competing males. For animals (cheetah, wolf, bear, etc.) who do not live in an organized social nucleus this is done after savage and deadly fighting. For those animals (lion) who live in a small social nucleus, the sexual problem may be simpler, but there are daily occurrences of invasive attempts by outsiders attracted by female scent. When there is more than one male, as happens in some primate species, a strict hierarchy is established based on strength. High-ranking males take the best females (females in estrus during the mating season), while low-ranking males take those left at the end of, or beyond, the mating season.

Sometimes the female in estrus may give the appearance of approaching a male, even taking a lordosis (dorsal inward curvature) position which may appear to be a sexual invitation. For species in a social nucleus, the female learns to comply. The lioness may be penetrated ten times a day very casually, and the female baboon may endure biting and beating if she does not cooperate. A practice seen especially in animals living in herds is the male continuously following the female without using too much violence against her until the annoyed female gives in. Through this method, the ewe is mounted by a great number of males during the ramming season. This technique of trailing the female is also practiced by the male cat.

Although most females cooperate (usually out of fear), their cooperation is not necessary. Depending on the animal species, different methods are used by the male to immobilize the female and secure penetration. In the mammalian world any combination of the following three methods is used: neck biting, placing the front limbs of the male on each side of the female chest to partially choke and immobilize her, and putting all the weight of the male body over the hind back of the female body.

When the female is as strong as the male, which is usually the case, the female not in estrus may not allow mounting, and if mounted will place the powerful proximal part of the tail against the vaginal entrance to prevent penetration. A male dog attempting to mount a bitch not in estrus (or even in estrus) may be sorry, because a powerful vaginal spasm may catch his penis and he may have trouble dislodging it. Such a defense mechanism is widespread among canidae, and certainly practiced by other species. In general, however, in spite of the discomfort or pain when the female is not in estrus, she usually will comply with the male. Courting the female is, therefore, unnecessary, and using that term for the events preceding penetration is a distortion of the word "courtship" and suggests a false similarity between animals and humans. What is observed is rudimentary and consists mostly of the male evaluating the priming (estrus) of the female. It is seen only in animals who are isolated and never in those forming a social nucleus. This remarkable cooperation results from the female watching the deadly fights between males and thereby realizing what she may go through if she objects, and at the slightest sign from the male she takes the lordosis position.

All scientists who report on the sexual act of animals have noted the lack of privacy associated with it. The animal sexual act is done publicly in front of adults of both sexes and of the litter, and no individual is sexually aroused, attracted, or repulsed by watching it. Animals may desire privacy in other functions: for instance, a cat always hides when having a bowel movement and covers his droppings, yet, he need not perform sex in private. Similarly, there is no sexual modesty or shyness in animals, and no attempt is made to hide or expose the sexual organs.

Sexual Position

The animal position of mounting or "*more ferarum*" is standard. The female stands on her four limbs, sometimes slightly spreading the fore and hind limbs for a better equilibrium and a wider base of support because of the weight on her back. The male climbs on the buttocks of the female with his forelimbs slightly spread on each side of her lower ribs, while trying to secure equilibrium on his hind limbs which are behind her buttocks. In spite of all the varieties found in the animal world, this sexual position is remarkably uniform. It is rather clumsy and it is true that, in most mammals, the anatomy of the body does not allow much variation. The variation of animals (whales) who face each other during the sexual act is rare.

During mounting, the female's concern is about how to maintain her balance and endure penetration without much discomfort or pain. The front limbs of the male squeeze her lower ribs on each side, possibly interfering with her breathing, and this pain will increase if she moves too much. She cooperates to avoid male violence and to make the sexual act as short as possible.

The concern of the animal male is different. He has to completely immobilize the female to perform the rather delicate and difficult penetration. He also has to protect himself from possible biting by the female if penetration is painful. If necessary he can bite her with little chance of being bitten back, since he is not facing the female. Some males (stallion or tiger) do not hesitate to bite the back of the female's neck and hold it firmly with their jaws. This neck-gripping technique is routine with the ferret and appears to be common to many species.

Sexual Act

Fortunately for both the male and female the sexual act is very short, a fundamental characteristic of sexual life in animals. It takes only one or two (three at the most) intromissions (penetration of the penis into the vagina) for ejaculation to occur. Complete penetration is not always achieved, and is not necessary because complete or incomplete presence of the penis in the vagina does not last more than a couple of seconds. If penetration lasted longer, the female would dislodge herself and a fight would occur. The speed with which the male performs the few successive intromissions attests to the fact that the animal is in a very uncomfortable position and eager to finish it. Sometimes ejaculation occurs as soon as the penis has touched the female genital area, before any real penetration. By human standards, this sexual act would be called premature ejaculation. The shortness of their sexual act is very understandable because at the time of mounting the secondary sexual organs already are fully distended with sperm and erection is already present. It is not the initiation or the performance of the sexual act that produces erection and filling of the secondary organs; if they were not fully distended with sperm, the male would not look for mounting.

A few, or even one good penetration of short duration is all it takes to trigger ovulation and relieve the female from estrus for the whole season. However, since many penetrations are incomplete, multiple mountings, during the same or successive days by the same or different sex partners are often necessary for coital stimulation to result in ovulation and estrus relief.

The trouble that many species experience with proper penetration is noteworthy. Before the tip of the penis finally finds the vaginal orifice, many males already have performed numerous but ineffective pelvic thrusts with the erect penis moving back and forth in the air. This ineffective search gives the impression of prolonged copulation with multiple thrusts; it is never the case. The rabbit, for instance, spends more time in ineffective thrusting, while only one single effective thrust in the vagina is needed to reach ejaculation. This applies not only to the rabbit but to all mammals, almost without exception. Most mammals have no chance of improving this situation since they do not have the use of hands to guide the penis into the vagina as humans do. Although many nonhuman primates masturbate and are, therefore, familiar with holding an erect penis, one never sees a monkey guiding his erect penis with his hand to make penetration easier.

Some domesticated animals, such as the bull or the stallion, would, if not for human interference, have the same difficulty in penetration as the corresponding wild species such as the buffalo or mustang.

Erotogenic Zone of the Animal Body

Contrary to humans, there is no sexual stimulation in animals through touching; no part of the animal body is erotogenic. Female body contact will not sexually stimulate the male, nor will male body contact lead to sexual stimulation or estrus in the female. When the male approaches the female for copulation, he already has been sexually aroused; the female rarely approaches a male sexually even when in estrus. Only the scent of a female in estrus is erotogenic. Physical contact does not play a role in sexual attraction or stimulation. There is no extra-genital erotogenic sensitivity in the animal body; no clitoral stimulation. Body contact in animals during sex is limited solely to what is required for proper mounting and immobilization of the female. The most extensive contact during mounting in animals includes not more than 10-15% of total body surface.

In many animal species, body contact is even more limited. The lioness, for instance, is very cooperative because she has to deal with a sexually demanding male; to make penetration easier for the male, she flexes her hind legs partly and the penis slips in easily. Body contact in this species is limited solely to penetration and penis-vagina contact. It is obvious that the two sexual partners do not look forward to any body contact before, during, or after sex.

Experimental sexual stimulation in animals is, therefore, done by means other than skin or mucosa. Mostly it is done by testosterone injection in males, estrogen injection in females, or pituitary hormones in both. Other techniques are electric or chemical stimulation of certain zones of the midbrain, which are preferred to the natural sexual cycle because these artificial cycles can be better controlled. Other techniques, such as mechanical manipulation of the genital organs, give limited or no results; it would be easier to control sex in animals by olfactory stimuli.

These observations lead to the bizarre, yet obvious, conclusion that masturbating the penis or touching the vagina-clitoris outside the mating period will in no way produce sexual stimulation of any kind; all that will be produced is a skin or mucosa irritation. Exceptions are made for animals grouped by humans in specific locations (cattle farms, zoos, etc.) where the pheromone stimulation can be continuous.

Emotional Aspect and Sexual Orgasm in Animals

Modern video cameras with powerful zoom lens have recorded the sexual act of almost all animal species in their natural environment. It requires a tremendous amount of scrutiny and courage to make certain statements concerning animal sexuality. First of all, it has never been demonstrated that animals have a sexual orgasm. It takes very little observation to conclude that the animal female does not have any orgasm, as human females do. Even under the closest scrutiny, using the most sophisticated equipment, nothing unusual between the male or the female during the sexual act is recorded. The face and bodies of the sex partners are expressionless; in fact, except for a few back and forth motions of the male body, there is barely any clue that the animals are having sex, or that ejaculation occurs. A male is known to have ejaculated only when he dismounts with sperm leaking. Limited grimacing can sometimes be observed in primates or other mammals, but it is short and not frequent. It is true that the facial musculature in animals is limited compared with humans, and that only limited facial expressions are possible. Other limited body motions such as yawning, tongue flicking, chin resting, and others have been reported in different animal species. Some authors have even described tachycardia (increased heart beats) and uterine contractions during ejaculation. If researchers looked for gestures accompanying other functions such as eating, urinating, defecating, and sleeping, they would find many of them. These are all very rudimentary compared to what is observed in human orgasm. The tail of certain species, like the dog, can be expressive of some emotion; yet it fails to register any specific emotion that could be indicative of something unusual even at the time of ejaculation. Both sexual partners are silent, and no excessive panting or special noise can be seen or heard in animals having sex.

The personality of animals is not expressed during the sexual act. Different dogs for instance, can have different personality characteristics (gentleness, aggression, loyalty, intelligence), displayed in their general behavior. Yet they are not expressed in the sexual act. Every breed of dog in the world performs the sexual act identically. Even if there were any traces of emotion during the sexual act, the sexual partners are not aware of it since they are not facing each other.

It is difficult to think there could be any possibility for orgasm in the animal female when sex is associated with fear and submission, when genitals are painful because of estrus and even more painful when not in estrus because penetration is more difficult, with the heavy weight of the male during mounting, with neck biting, and most of all with almost instantaneous ejaculation. A human female would never achieve sexual climax under similar circumstances, i.e., rape. The behavior of the human female during sexual climax is characteristic of intense emotion, and no matter how extensive is their imagination, observers will never remotely see in the animal female any characteristics of the sexual orgasm of the human female. The animal female is so silent and immobile during copulation that it is impossible to imagine she had an orgasm.

Similar observations and conclusions can be made about the animal male. There is no sexual orgasm as in the human male. One could argue that ejaculation is always associated with a male climax. This is just a human concept transferred to animals, and even in men there are many exceptions. Men who have had a prostatectomy can reach orgasm without ejaculation, and men suffering from premature ejaculation can have an ejaculation without an orgasm. In the human male, the quality of the orgasm is in relation to the quality and length of the coitus. Generally, a very short coitus is associated with a male orgasm of poor quality, if any, while the best orgasm requires an optimum coital duration. Since animal coitus is short, almost as short as human premature ejaculation, it is also logical to assume that animal male coitus is not associated with orgasm. One never sees any panting, excitement, intense emotion, screaming, convulsion-like movements, then resolution phase, as seen in the human male and female. After ejaculation all one sees is an animal male walking away obviously relieved because the distended seminal vesicles are emptied.

After observing a variety of different mammals performing sex in zoos and shown in scientific programs, I came to the conclusion that what these animal males experience is not a sexual orgasm of the nature described in humans but relief coming from the evacuation of painfully distended seminal vesicles, similar to the relief from evacuation of any distended organ such as bladder or rectum. In my opinion this is the only orgasm or pleasure the animal experiences from sex. Observing a lion during coitus is no different from observing a lion evacuating his bladder or rectum. In fact, if one could block the female from view, it would be difficult to know if the lion was defecating, urinating, or ejaculating; his attitude before, during, and after any of these three functions is identical. The fact that the seminal vesicles are already filled at the beginning of the sexual act in animals is what leads to a rapid ejaculation.

Coitus is the only way to relieve painful male and female estrus, otherwise animals would never pursue sex. Human female climax frequently occurs after male climax, necessitating further male thrusts after ejaculation; yet, an animal is never observed making additional thrusts after ejaculation.

Sex at Different Stages of Animal Life

The different periods of animal sexual life such as prepuberty, early puberty, full sexual maturity, pre- and post- menopause are perceived by animals through scent and have only an olfactory name. Prior to puberty there is no awareness of sexual differences. At that stage it is unlikely that parents or the litter know that the cubs are of two different sexes. There is no sexual interest whatsoever.

Sexual education in animals is almost nil and certainly takes less time than learning to hunt for food. Mostly, there is play either among the young ones, or among other young ones and their mother. They may touch each other's genital organs and there even may be some mounting, but mostly in terms of general touch rather than with any specifically sexual intention. It is possible that during prepuberty there is a certain amount of play mounting between siblings or with the mother, either initiated by arising puberty or simply copying the parents. At any rate since the sexual act in animals is very short, there is not much to learn.

Puberty of the litter always means disruption of the family unit and, rather than sex, the concern of the rejected adolescent is how to survive. The animal adolescent female becomes a sexual competitor for the mother or the adolescent males are expelled by the adult male. In species like the baboon or the lion, who form social groups, the sexual advances of the young males are resented by one or two senior males of the group. There is, therefore, no opportunity for sexual education or practice.

The outcome for the adolescent female is sometimes different in these groups. In this case, female sexuality starts immediately at puberty. In early puberty, or even before, the young female may undergo continuous rape attempts, and the male usually gives up only because there is no room for satisfactory penetration, or it is too painful for the young female who may fight back. However, at the earliest sign of puberty, the young female is forced into a sexual act by a strong male which easily could be the father, even if her body is not mature enough to carry a pregnancy. The lion, for instance, will casually add her to his own pride.

Immediately at puberty, the young animal male is capable and desirous of sex. The search for a female in estrus may begin in adolescence, and she is easily sniffed miles away. Unfortunately, most of the time, many adult males get the same "sniffed" message, and when the female is reached, other suitors are already there. Animals quickly size up each other's strength and many of the wooers withdraw voluntarily or are quickly eliminated. There is no room for adolescents. If there is any serious or real competition it will be limited to the two strongest males. The real sex education for a male is to learn how to fight for a female, and not how to court a female or how to perform sex. For animals like the lion living in a social group, the fight is even more serious because more is at stake. The head of the pride will put his life on the line rather than be displaced. The reward is great if the owner is displaced or killed. The new tenant immediately owns many females, with which he can indulge continuously. The reward is even greater because a hunting territory goes with the pride as well. The new problem now is not to be eliminated by a stronger male from the outside.

During the adult life of those individuals who indulge in sex, the sex life is monotonously repetitious. In fact, the way animals practice sex is the way they live in general. Their sex life may be regulated by genetics, instinct, observing, and learning from other animals. No matter how it is acquired, it is done the same way, and will be repeated by the animal for life. The animal repeats the rudimentary sexual act as performed by its parents and is not expected to innovate.

Sexual menopause is rare in wild animals because they usually die before menopause occurs. The little that is known about advanced age in animals applies only to those animals whose life is regulated by humans. Their way of life is not governed by Darwinian evolution and, therefore, they may live much longer than in the wild and may go through menopause. With the hormonal changes of menopause, they produce no specific scent and the female loses her sexual identity. She is no longer approached on a sexual basis. The male will no longer produce sperm, and with no distended seminal vesicles or erection, he loses interest in sex. At any rate, post menopausal life span in animals is either not existent or very short.

Sexual Deprivation in Animals

The most important aspect of male animal sexuality is that it is reserved for the elite, the physically strong. All others are sexually deprived for all or part of their lives. Not all sea lions, lions, or baboons can find a group of females at their disposal. Only the strongest male can conquer, and he takes as many females as he can handle. It is only during this short period of his life that he remains the strongest. He will be ruthlessly displaced at the slightest sign of weakness. Even in animals (bears) who do not live in groups, it is likely that the female's sexual favors are reserved only for the strongest males who can impose themselves physically. This sexual deprivation is serious for animal males since their sexual need, depending on the species, can go up to ten or fifteen sexual acts a day during the mating season.

These sexually deprived males must suffer markedly from the distention of the secondary sexual organs, but eventually these vesicles get emptied spontaneously in different ways. They can empty progressively and continuously, and the content is eliminated and mixed with urine. The sperm also can be eliminated at night during sleep, commonly called nocturnal ejaculation. Another form, most likely very common among animals, is spontaneous ejaculation done while the animal is awake. The animal male who has no sexual partners learns to empty his seminal vesicles when they are filled, the same way he empties his bladder or rectum. Since the sexual act in animals is very short, it takes a minimal amount of rubbing the skin of the penis against the vaginal mucosa to produce ejaculation. The likelihood, therefore, is that minimal rubbing of the erected penis against any object will produce immediate ejaculation and relief from the distention of the seminal vesicles and from erection of the penis. It is well known that male cats do it all the time, and the likelihood is that it generally occurs in the animal world. The veterinarian who masturbates the bull to get his sperm for artificial insemination knows well how quickly and easily ejaculation occurs.

After two animal males have fought for a female, sometimes savagely, one of them gives up and runs away. Usually it is because he realizes that he is weaker, but I suspect there is another reason. During the fight there is a tremendous amount of adrenaline secretion which could lead to contraction of the seminal vesicles, vasa deferentia, and prostate and produce ejaculation in one of the two males. Since penetration is no longer necessary after this spontaneous ejaculation, one walks away completely relieved.

Another form of relief is homosexuality. In humans, male homosexuality is the full intromission of the penis into the anal orifice of another male; this is preferred to heterosexuality. Such practice is nonexistent in animals. Some kind of vague homosexual play can be observed as a part of general play, mostly in young ones. Real anal penetration would be too painful and might produce a violent reaction on the part of the mounted partner. What is seen, however, mostly among primates but also in lower animals, is rubbing of the erected penis in the anal or perianal area. Most of the time penetration is unsuccessful because of dismounting. Occasionally, if the position can be maintained for a few seconds, ejaculation will occur since it takes very little rubbing of the penis to elicit ejaculation.

It is also possible for the animal female in estrus not to get sexual relief if she produces less pheromones, or for any other reason. If she is not approached by a male, then she has to go through the discomfort, sometimes painful, until congestion of the genital organs recedes spontaneously as the estrogen level recedes. This may take a few days or weeks. This occurs occasionally in groups formed by primates when the male population gets smaller, or when some females are less attractive olfactorily, than others. In general, animal females are better off in a harem under the protection of a strong male rather than subject to the haphazard sexual requests of casual males. In a harem enclosure, they can at least share the sexual requests of only one very demanding male while outside a harem they may be subjected to frequently repeated and possibly dangerous requests from many males. My impression is that the animal females look forward to protection and this may be the reason they readily join and remain in a harem.

Incest in Animals

It is not my intention to discuss the meaning of incest here. This taboo, which has been discussed by many authors, actually exists only in humans. It would be wrong to state that animals practice incest the way humans do. It would be more accurate to state that the concept of incest (a taboo) does not exist in animals, since taboos do not exist in the animal world. The animal does not know the difference, sexually speaking, between his family unit and the outside world. He looks for sex to occur the easiest way, and if it happens to be easier in the family unit, it is done so, incestuously. The lion has sex with his daughter not because he wants to rape his daughter (a human concept); he does not even know which among the young ones is his daughter. He just has sex and does not know how to make the distinction, or rather he does not know there is one. Similarly, the young male is having sex with his mother, not necessarily because he wants her specifically but because he is more familiar with her body and because she may be more tolerant; if any other female would show the same tolerance, which is highly unlikely, he would indifferently switch. There is really no incest in animals, only the inability to make a distinction between sex within and outside the family unit.

Isolated Exceptions to the General Rules

While these characteristics are general to all animals, there are many exceptions. Some species, especially among primates, may show rudimentary trends to get away from the monotony. Foot clasp is practiced by some monkeys. Rudimentary forms of orgasm, most of the time detected only by special equipment, but not really evident visually, have been reported. Some primate females may have a non-seasonal estrus and menstrual cycle. Monogamy, a rare occurrence, is observed in the gibbon, the marmoset monkey, and some other primates, at least on a temporary basis. Male aggression may not always be manifest, and high-ranking males may show some tolerance to low-ranking males. Other monkeys do not necessarily break up the family unit at the onset of puberty of the litter. In some gorillas, sexual behavior can be somewhat moderated by environmental and social factors. Penetration may go beyond a few seconds, especially when the male is exhausted from a preceding fight, as with the bear.

Sex in chimpanzees has been described in detail by Jane Goodall. A female chimpanzee is in heat; the males form a line determined by hierarchy and mount her successively. The female barely looks at the successive sexual partners and Goodall (1971) wonders why it is that a specific female is preferred to other females in heat, since the female selected was comparatively old. My suspicion is that the scent of pheromones coming from the older female is stronger than those of other females in heat; and that the concept of human beauty, based on physical appearance and youth, does not exist in animals. It would be pure speculation to imagine it in animals.

The most important exception, not by its frequency but by the debates that result from it, is the occurrence of front-to-front, ventral-ventral, or face-to-face copulation in animals, the way it occurs in humans. Some field researchers have come up with personal observations reporting ventral-ventral copulation among pongids and other primates. There is no doubt that some of these observations are genuine. Although I never explored the jungles of South America, Africa, or South East Asia, for years I have been videorecording television programs on behavior (including sexual behavior) in different animals and specifically in primates. I have yet to see a videotape of front-to-front sex in any animal. However, given the fact that reliable researchers have reported it in animals, and produced photographs, it must be considered, even though it is rare.

Why some researchers have reported ventral-ventral sex in animals may have to do with general animal behavior. One difference between animals and humans as far as sex is concerned, is that the human body is usually divided into sexual areas (breast, perigenital) and non-sexual areas (the rest of the body). In humans, any hand contact by the male partner with the breast or genital area of the female partner, or any posture that can be in any way reminiscent of sexual posture, is immediately interpreted as sexual in nature. Animals, especially primates, play and fight together continuously, touch each other (romping, grooming), and position their body in relation to playmates in all kinds of postures. This is understandable since their four limbs function as either four hands and/or four feet, with all kinds of positions possible. It would not be unusual for an individual to roll on its back and the playmate to jump on him or her in a posture that our human norms interpret as sexual. After watching numerous scientific programs on television and observing groups of monkeys, baboons, gibbons, and chimpanzees in zoos or in jungle habitats, I have seen positions taken by these primates which, if observed or filmed by humans, definitely could be considered sexual in nature (although the animal was not in heat and had no sexual intentions whatsoever).

The other exception concerns permanent mating in animals, something akin to marriage. The classic example is represented by the gibbons who supposedly mate for life. Further field studies have shown, however, that there is a question of supply and demand, that males and females are spread out more or less evenly, and even though it is true there is usually one sexual partner at a time, it is not the same successively, because the partner changes frequently. It should be added that gibbons are very aggressive and hyperactive animals (males as well as females) that never stay in one place, and securing a harem of females might be a problem for even the strongest male. Finally, bonding, a special relationship between two individuals that goes beyond the sexual act, has occasionally been described in some primates. This subject will be discussed later in detail.

These "human" trends are very isolated, superficial, short-lived, and often exist only in the imagination of the observer. Most of the time even the highest primates move back and forth into the general characteristics previously presented. Most of these exceptions are seen in the artificial conditions of zoos and jungle habitats. They are seen also in animals tamed by highly skilled and very knowledgeable animal psychologists. This is very rudimentary, however, and as soon as the animals return to the wild, they revert to their original patterns.

In summary, the law of the jungle applies to sex as to every aspect of animal life.

HUMAN SEXUALITY

Introduction

Much has been published today on human sexuality about anatomy, biology, psychology of sex, and its deviations being described in detail. Much of the general literature on sexual techniques claims the secret of perfect sex. Conferences on human sexuality are well attended. Sexual stories and scandals in newspapers and television news are always well read and watched. The only innovation presented here is to compare various aspects of human sexuality with animal sexuality, i.e., courtship, sexual positions, ejaculation, etc., with an aim to shed some light on what is animal and what is human in sexuality. Therefore, only the characteristics of human sexuality that are essentially different from animal sexuality will be included.

Cerebral Versus Biological Control in Human Sexuality

The same sex hormones present in animals are in human males and females, but they work differently, as follows:

(1) The hormonal cycle in the human female functions continuously and cyclically on a monthly basis from puberty to menopause, while in animals it functions only seasonally during the mating period.

(2) In the human species, the estrus (or painful swelling of the perigenital area) is weak or even nonexistent. It is called concealed estrus. Cytological changes in the vagina and histological changes in the endometrium can be detected, but the obvious behavioral changes so characteristic of the animal female estrus are not seen. The production levels of different hormones can be shown in blood or urine tests but rarely are any marked changes in the sexual behavior of the human female seen, which could, even remotely, be similar to the behavioral estrus of the animal female. Specifically, human estrus is painless and usually goes

unnoticed.

(3) While ovulation in animals occurs mostly after coital stimulation, in humans it always occurs spontaneously at midcycle, with or without coitus. If no impregnation occurs, the woman will pass the egg even without intercourse and menstruate two weeks later. Sexual intercourse in the animal makes the ovary pass an egg, and this eventually leads to pregnancy.

(4) Male stimulation is not produced by any pheromonal sniffing. There is no pheromone production and no scent indicative of the female partner's receptiveness. The animal type of male stimulation does not exist in humans because pheromones have disappeared, making male sexual stimulation completely different.

(5) Teleologically speaking, sexual behavior in animals is geared solely toward impregnation of the female and reproduction; this is the whole purpose of animal coitus. In humans, sexuality can be unrelated to and separate from fertilization and has other purposes besides reproduction.

(6) There are some exceptions, however. A weak but significant estrus may be detected in some women and associated with some definite psychological changes. Human coitus may lead to ovulation, although this is rare. Women may be psychologically predisposed to sex during ovulation. All these exceptions are rare and unpredictable and not of the intensity observed in animals.

Human sexual activity certainly supposes normal hormonal levels since diseases associated with abnormal sexual hormone levels frequently are associated with sexual disturbances. Therefore, sexual hormones play a definite role in human sexual behavior, but they do not rigidly determine and control it the way they do in animals. Animal sexuality can be predicted by following the hormonal variations in blood or urine, or just by observing the biological estrus in the male and female. Human sexuality is not entirely biological as it is in animals.

Role of the Higher Cerebral Centers in Human Sexuality

If human sexuality is not completely and strictly controlled by hormones, as in animals, then what controls it and how is it regulated? It is hard to answer this question without controversy, which I want to avoid. There is no doubt that the higher cerebral centers, so developed in humans, exert an important control. The appearance of the brain as the center of all operations is an evolutionary process culminating in man. Some simple situations demonstrate the role of the higher centers of the brain in human sexual behavior, as follows.

(1) A young man sees an x-rated movie and is sexually aroused and has an erection; yet, this young man will not rape the first girl he sees. A young lady reads a book on sexual techniques and becomes sexually aroused and very desirous for sex; yet, she will not propose sex to the first man she meets. Such sexual behavior would be considered highly unusual for the average human being, and yet this is the rule among animals. Humans can exert full *conscious* control over their sexual impulses, while animals are completely controlled by them. Although both men and women have sexual desires, it is a common observation that it is easier for women than men to control their desires.

(2) Two women may have the same estrogen levels and cycles and yet one is very inhibited sexually while the other is promiscuous. Two men may have the same testosterone levels and yet one is a family man having intercourse with his wife once or twice a week, while the other is living the life of a playboy. These examples show that *unconscious* psychological factors are of some importance in determining human sexuality. We do not know how many unconscious factors there are in animals but, even if they exist, they do not appear to play a major role in sexual life. It is not the purpose of this presentation to discuss the nature of the unconscious, whether it is composed of repressed memories (Freud, 1930) or whether it is also influenced by archetypes (Jung, 1957b).

(3) Without prior biological stimulation a man and a woman who meet and are attracted to each other can become involved sexually. It could be a husband and wife in their bedroom, or a boy-meets-girl at a casual party. In either case, neither one was sexually aroused when they met nor had any sexual intentions at first, but they will work themselves and each other up to a state of sexual stimulation, eventually having sexual relations.

Some conclusions can be drawn from these examples. Biological factors are not dominant in controlling human sexual behavior as in animals, because the effects of sexual hormones (male and female) are easily overpowered by highly developed brain centers.

Higher cerebral centers, symbolizing the personal and social background of the individual, play a fundamental role in determining human sexual behavior. Sexual hormones do not automatically determine sexual action; the *conscious* mind has to clear it first or the *unconscious* mind may interfere.

Without any prior biological stimulation, men and women can create or initiate their own sexual stimulation ex vacuo (without any prior intention) at any time and under any circumstances they wish.

Of course to understand this better, I have accentuated the differences between humans and animals. Actually, control is not always easy and very often men and women give into sexual impulses. Although there is a continuous struggle for power between the conscious and the unconscious factors, and biological stimuli, a variable amount of conscious control is always possible. At any rate control of the sexual impulse is the fundamental difference between human and animal sexual behavior, with humans, not animals, responsible and accountable for their sexual behavior. Human sex is as much cerebral as it is biological, and there is no doubt, therefore, that the original site of sexual stimulation is not the same in animals as in humans.

In animals, sexual stimulation originates in the midbrain (or biological brain). The order is imperative, absolute, and transmitted from the midbrain to the conscious brain for immediate execution. The cortical brain does not decide whether or not to have sex, it just receives the order to carry it out.

It is different in humans. Under hormonal stress, the midbrain may occasionally send sexual signals to the conscious brain, which in no way is mandated to carry out the order. The final decision to engage in, look for, or inhibit sex is a conscious and voluntary decision. Besides, the conscious brain can receive sexual stimulations directly through tactile, visual, and auditory senses, or it simply initiates its own intrapsychic stimulations. Therefore, between animals and humans, totally different pathways are utilized in terms of sexual stimulation.

In fact, the different nature of primary sexual stimulation explains why the degree of sexual control exerted by humans varies from one individual to another. All the intermediary steps can be seen between the rapist on one extreme and the sexually inhibited on the other. Their biological sexuality is similar, but their cerebral conscious sexuality is different.

Since the hormonal role is not primordial in humans, sexual activity, or at least sexual interest, can be seen in all phases of human life (before puberty and after menopause) even when hormonal activity is not present.

Variety in Human Sexuality

Because the conscious and unconscious mind is different from one human individual to another, there is a tremendous variety of sexual patterns in men and women, depending on the individuals and the different stages of their life, social level, and culture. Human sexuality is difficult to describe because, compared with animal sexuality, it becomes obvious that not all human characteristics are present in human sex and not all animal characteristics have disappeared entirely from human sex. Therefore, not all human sexual acts will completely fit the description that follows; some characteristics are more common than others.

I will start with the human sexual position. *More hominum*, the ventral-ventral position, is specific to humans compared with *more ferarum*, or ventral-dorsal position (mounting), in animals. Where humans are really different though is in the infinite variety of sexual positions taken before, during, and after the sexual act. The animal position is one among many that are described for humans. Each human couple manages to find the best position suitable to individual personality, culture, and intimate (or lack of) relationship.

In animals, not only is the sexual act short in general, but it is fixed for each species, with the rabbit having only one intromission and the bear having up to ten. In humans it can be as short as in animals or it can last one hour or more.

Not only is there marked variety in human sex, but it seems the variety is still expanding. Men and women are not necessarily bound by a learned pattern. They are free to add their personal innovations to the already numerous existing patterns. Variety rather than monotony is characteristic of human sexuality.

Role of Sight in Human Sexuality

Humans have lost estrus (main stimulus for female animal sex) and the scent of pheromones (surest stimulus for male animal sex and indication of female sexual readiness). I have shown that humans have an increased sense of vision, hearing, touch; scent and taste are less developed. The logical conclusion is that these new functions replace estrus and pheromones and play an important role in human sexuality.

I begin with visual stimulation. Ninety percent of the information stored in the human brain arrives through the eyes, the rest through hearing and touch. Information derived from sniffing and licking is almost nonexistent. Each individual attempts to create for himself and others a visual attraction, i.e., the equivalent of visual estrus or pheromones. Let us describe first what women do to reinforce their external feminine appearance. "Visual" pheromones have been substituted for the lost olfactory stimulation by pheromones. The conscious maneuvering of the visual characteristics result in what could be called quaternary sexual characteristics (the other characteristics are anatomical and have been previously described in a previous chapter. Enumerated below are methods women use to reinforce their external appearance:

1. The height of the body is altered by high- or low-heeled shoes.

2. Hair, which can be dyed, curled, styled, ornamented, is not necessarily longer on the head; eyebrows and eyelashes are altered; hair on the lower limbs and the perigenital areas is removed.

3. An excessive muscular appearance is avoided by careful physical exercise.

4. Fat distribution is controlled by special girdles, panties, and stockings.

5. Breast shape is carefully controlled with special bras.

6. The art of make-up or cosmetology.

7. The art of wearing status-item clothing, which represents a large portion of a woman's and even a man's budget.

8. Body massage, while of doubtful value, is very popular for manipulating body appearance.

9. Plastic surgery and modern medicine to enhance beauty and extend youthful appearance.

10. Miracle lotions, vitamins, hormones, drugs, etc., although worthless and sometimes even dangerous, used more as psychological props.

Darwin thought of sexual selection and beauty primarily in terms of ornamentation such as tattooing, lipstick, high-heel shoes, bouffant hair styles, long neck, and all the other ways humans have of altering their appearance.

This long list of external reinforcements should in no way give the impression these manipulations are exclusively feminine. In our present culture, men are also involved with these manipulations, although to a lesser degree. They, too, pay attention to their hair, beard, mustache, body build, clothing, etc. Men base their sexual attraction also on business or intellectual accomplishments. As noted by a Chinese philosopher, the whole behavior of Western man is geared toward attracting women (Waley, 1964).

One day while interviewing a three-and-a-half year old boy, I asked if he was aware of any difference between a boy and a girl. His answer was: "Boys have a penis, girls have makeup." To this day, I still wonder about the difference or rather the similarity between a penis and makeup.

These visual manipulations are purely individual and cultural. There are or have been cultures where these manipulations were mostly masculine. The American Indian braves of the previous generations extensively decorated their bodies with tattoos and ornaments, while the squaws barely used any. In the royal courts of the 17th and 18th centuries, men used clothing, makeup, and wigs that would easily match the feminine mode. Flamboyant military uniforms also were devised for visual attraction.

The perception obtained from this is that the whole human body becomes sexual, while in animals there is only a small area. Sexuality now involves the whole body that is sexually active all year round and uses all kinds of visual, auditory, and tactile stimuli. There are no more neutral zones in the male and female bodies. Indeed, the human body quickly lost its neutral characteristics and every part of the body has become linked with sexuality, although clothing, hair, etc., do have functions other than sex. This is more obvious in the female, but similar observations can be made about the male body, although to a lesser degree.

The importance of sight in human sexuality is expressed by the sense of esthetics or visual beauty. Facial musculature in humans is very developed with 43 different muscles of varying lengths and strengths that can contract at different degrees and in different combinations. This role will be discussed later.

Role of Touch in Human Sexuality

We have tactile estrus and tactile pheromones. While vision plays an important role in the selection of the sexual partner, touch is crucial for sexual stimulation and arousal. Some parts of the human body are more sexually stimulated by contact than others, but there is practically no part that is not sexually stimulated by contact. During coitus almost every part of the body of one sexual partner comes in contact with every part of the other's body. The new discoveries of the human body are the lips for kissing and the breasts and the clitoris and male genital organs for touching. These contact stimulations are important and necessary in humans, while in animals they are nonexistent and unnecessary. The presence of clothing during human sex may restrict but does not eliminate contact stimulation.

Although most animals like to be touched, continuous touching for any length of time will never result in sexual arousal. Similar touching in humans would most likely end in sexual arousal.

Similar remarks can be made about masturbation. In humans, this practice will bring a penis from a state of nonerection to full erection and ejaculation without any prior desire for sex. Similarly, clitoridal masturbation will result in female orgasm. In animals, it is different: contact stimulation of a nonerected penis will never bring erection unless the animal is under hormonal stimulation and ready for erection. Only humans can bring themselves to a state of sexual stimulation by contact manipulation.

What has been said about visual and tactile stimulation also can be said about auditory estrus and auditory pheromones: the human voice and intonation can be sexually attractive. Both men and women indulge in it, although it is more obvious in women. Vocalizing can be part of the sex act, even a main part, as in phone sex.

Role of Scent and Taste

Sexually speaking, scent and taste have almost disappeared in humans. Normal body odors do not play much of a role in human sex; in fact, they appear to have counter effects. Body washing is certainly healthy, but when done excessively it rids the body of natural odors. A clean smell is nothing more than the elimination of natural body odor.

In humans, smell is not a mode of communication as it is in the nonhuman mammalian world. There are a few remnants of animal physiology left in humans, and olfaction is encountered slightly more often than taste.

However, there is an on going attempt on the part of the human species to revive the dying olfactory function, at least from the sexual point of view. This attempt has been well described by Forsyth (1966) and Ackerman (1990). For that reason, women and occasionally men generously use perfumes. The goal is an attempt to take advantage of a dying sexual stimulus that flourishes so well in animals.

In humans, tasting is not a mode of communication the way it is in animals. People do not taste or lick each other, except in the most intimate mode of relationships, such as a sexual one. Aside from this specific situation, people who constantly lick each other the way animals do, would be considered highly abnormal and even pathologically ill.

In summary, scent and taste have markedly weakened and are being replaced by sight, hearing, and touch in human sex. The powerful, but short, stimulation by olfactory pheromones is replaced by the mild, but constant, stimulation of visual, tactile, and auditory origins.

Last but not least, human sexual stimulation can be initiated without any sensory stimulation, such as mental and emotional stimuli that can function as estrus or pheromones. In conclusion, initiation of sexual desire is very complex in humans and is an interaction of conscious, unconscious, sensory, and biological stimuli.

Individualization of Sexual Partner

An important characteristic of human sex is not just having sex, but sex with a specific partner. Men and women desire sex with a specific person, because they are not drawn to sex solely as a result of hormonal stimulation. The sexual partner is selected through visual, auditory, touch, or mental recognition, and it is the specific partner that is sexually stimulating. It is obvious that the higher brain center plays a fundamental role in this selection and individualization. However, it is important to stress that this is more a trend than a reality. There are times when men or women just want to have sex, but even in the extreme situation of a man going to a house of prostitution, some choice is exercised in visually selecting this temporary sexual partner. Women tend to individualize the sexual partner more than men, although both actually do. Individualization of the sexual partner affects human sex in different ways, as follows.

Uniqueness, a consequence of individualization, is very characteristic of human sexuality. No two women, or two men for that matter, want to look alike in their clothing, hair style, makeup, etc. Women want to be unique in appearance, as do men but to a lesser degree.

Intimacy is also a logical sequence of individualization. As sex becomes individualized and the two sex partners get to know each other more deeply, their sex life reaches new dimensions and their relationship expands beyond a sexual one, persisting even when sexual life has ended.

Sexual faithfulness and monogamy are the natural expressions of individualization. Men and women can have different sexual partners, but they are usually successive. Even in polygamy, most times there is a preferred partner (at least for some period of time). The only time there is indiscriminate use of sexual partners is in prostitution, and even then there is usually a preference for a particular partner or pimp. Fear of intimacy is one reason for promiscuity in humans.

Jealousy is another aspect of sexual individualization because substitutions are resented.

Human Sexual Approach

In humans, physical aggression is not necessary. Overpowering a woman physically is rare since her cooperation is usually necessary for the sexual act to take place. This does not make human sex easier. On the contrary, more refined qualities are required of the male in seducing the female and of the female in attracting the male. The human approach requires planning and time. Careful and skilled courtship secures female cooperation and male attraction.

In the human species, all the events preceding penetration of the penis into the vagina are necessary because it is mostly during that time that sexual arousal occurs. Contrary to animal sexual partners who are already in a state of sexual arousal when they meet, the human partners have to work themselves up to a state of arousal during a period of courtship and foreplay, then proceed with the sexual act; they touch each other all over, but mostly the woman's hands with the penis and the man's hands with the clitoris and the breasts. In summary, an elaborate and non-aggressive preparation to sex is characteristic of humans.

Sexual Act

It is during the sexual act itself that these differences become more obvious. During animal coitus, the concern of the female is to steady herself and to move as little as possible in order to make sex painless, while the male mounts and immobilizes the female to secure proper penetration.

In humans, the woman offers the front part of her body. She rests comfortably on her back, making it is easy for her to carry the weight of the male. A concerned male partner can support part of his weight on his knees and elbows, without the inconvenience of the woman totally supporting his weight, so that the front of the two bodies can fully touch. The hands are now free; the whole body is free for any manipulation. The face-to-face position (as revealing as the ventral-ventral position) has replaced the mounting position used by animals. Because humans are free they can concentrate on other goals. Not all these other goals are present each time humans have coitus. At one extreme, nothing may occur and sexual intercourse may be very rapid as in animals; at the other extreme, all the differences from animals that has been or will be enumerated occur. What happens during the human sexual act depends upon the relationship between the two sexual partners, their age, and state of mind at the time of coitus.

The features of the human sexual act are easy to describe if the important role of the higher cerebral centers in human sexuality is kept in mind. In animals, every step is regulated instinctively and biologically; in humans, each and every step of coitus is, or at least can be, under total conscious control. Man can make penetration brutal and painful like animals, or make it gentle and easy with the help of his hand and/or with the use of a lubricant; he can make the successive intromissions slow and deep, enjoyable, or make them superficial and very fast. Modern sexologists have pointed out the possibility of withholding impending ejaculation whenever desired, thus waiting for the woman to get closer to her own climax, as will be elaborated upon later. Women too may wish to end it fast, or to participate in one or all of the different steps of the sexual act. Gentle caresses replace the firm immobilizing grip of the animals; delicate patting replaces scratching and clawing. The mouth and lips become erotogenic and kissing all over the partner's body replaces animal biting or threat of biting. Biting, a remnant of male aggression or female defense, can still be observed in human sex.

By animal standards the human sexual act is very elaborate (Masters and Johnson, 1966; Kinsey, 1948, 1953). Intense emotional displays and excitement (sweating, panting, groaning, tonic-clonic phase), then the resolution phase, make human coitus more expansive than the precipitous penetration and ejaculation of the animal.

I have described the short duration of animal coitus. Researchers of animal behavior rarely describe animal coitus because there is nothing, or almost nothing, to describe! Human coitus is prolonged, with an average copulation time of 10 to 15 minutes.

The culmination of the sexual act, or sexual climax, is called orgasm. It varies tremendously in quality and intensity. In the human species, one should not confuse orgasm with ejaculation. Either one may occur without the other. When there is too much conscious or unconscious tension, neither orgasm nor ejaculation may occur.

Especially in human females, orgasm should never be taken for granted. Many women in the spectrum of modern, primitive, and past cultures go through life unaware of its existence. Even many men do not know, or do not want to know, that female orgasm exists and do not care to participate in it. As previously described, orgasm does not exist in the animal female. Orgasm in the human female is rarely spontaneous. To elicit it is an art, which is acquired during a continuous process of maturation, and even then, it is not that frequent. Women who believe they can reach an orgasm during a quick and casual sexual encounter, or the man who believes that he can elicit an orgasm in a female under similar conditions, live in a fantasy world. In women the orgasm is more complex because copulation is not usually associated with any physiological phenomenon such as ejaculation in men. Men after all have the advantage that ejaculation almost always triggers some kind of orgasm, but in women there is no biological phenomenon equivalent to ejaculation.

Orgasm in women has to be built ex vacuo, i.e., without any biological base whatsoever; the conditions for its occurrence are elaborate and delicate. It requires intimacy, time and privacy, peace of mind, individualization of the sex partner, gentleness, caring, love, patience, careful preparation, etc. Elaborate and delicate mechanical techniques acquired only after prolonged intimacy are necessary to elicit it. Some women can lead a normal sexual life (even motherhood) without ever having a sexual orgasm, while others may reach it in each coitus. The greatest error to be made in the study of human sexuality is to consider the female orgasm biological. This is completely erroneous for the simple reason that female orgasm does not fulfill any biological function and is not at all necessary for conception. *It is the greatest human acquisition.* Again, one can see that during the sexual act, biology is dominant in animals, but in humans it is markedly interfered with by the higher cerebral centers. In animals, proper hormonal levels guarantee erection and ejaculation, but in humans even though the biological hormones are necessary they play only a small part in the sexual act. Of all the problems associated with human sexuality, I never saw one cured by the administration of estrogen in a woman, or androsterone in a man.

How sexual orgasm originated in humans will be described later.

Persistent Relationship After the Sexual Act

Nothing is more disheartening for animal lovers than to observe the end of the sexual act in animals, or rather the attitude of the two sexual partners after the male has ejaculated. The female walks away obviously relieved because she was about to collapse under the weight of the male. The male, relieved after emptying his seminal vesicles, walks away in the opposite direction. Those who might argue about the intimacy, the togetherness, the close relationship between the animal male and female, are puzzled by how the two partners walk away from each other, like two strangers with nothing in common.

In humans, a similar termination of a boy-meets-girl casual encounter that starts and terminates quickly also can be observed. But this is rare and even then there are usually false promises of meeting again. Most of the time the intention is to make more out of it than there is. The relationship, however, can be of variable duration, or a prolonged and even permanent one. The intention is to consider the sexual act as an aspect of the whole human relationship rather than an isolated act by itself. Or, the intention of one or the other partner may be against any other relationship except a sexual one.

In animals there is no such occurrence. An animal female may get attached to her litter and a domestic animal may be attached to the human it lives with, but an animal is never seen developing any kind of relationship with the opposite sex just because they had coitus. It is just a biological act and creates no bond of any kind, neither love nor hate, thus making it different from the human one.

By now, the sexual differences between animals and humans are obvious. The animal male is stimulated by pheromones coming from any female of the same species (not necessarily by the female he is about to mount). This olfactory stimulation is temporary, violent, and sudden in its appearance and disappearance. In humans, sexual stimuli are milder in nature, but they constantly and continuously bathe males in an atmosphere of sexual stimulation which slowly but continuously fills their genital organs with sperm; they are practically always ready for a sexual act. However, because human stimuli are much milder, it takes days, or at least hours, for the genital organs to refill with sperm, unlike the powerful animal pheromones which act more quickly and can refill the genital organs within minutes. Therefore, the human sexual act lasts much longer than in animals, although it occurs much less frequently.

Beauty, or visual appearance of the whole body and specifically the face, is a sexual stimulant only in the human species because the visual sense (among others) has replaced the pheromones as a sexual stimulation. It will appear with adolescence as the human female body forms and will diminish in later life as folds and wrinkles of the skin appear. It is a common belief that appearance of the human female body is sexually attractive only as long as a youthful appearance is maintained. I will show later that this is not necessarily true. Similarly, the animal female body is sexually attractive only as long as intense (quantity and quality) pheromones are produced.

Now I can outline the fundamental differences between animal and human stimulation. The difference between humans and animals is the pre-ejaculation part of the human sexual act which forms the bulk of copulation in humans, and which is nonexistent in animals. The only similarity between animals and humans is the ejaculation phase, which comprises the whole animal sexual act and only the end of the human act. Therefore, it is easy for a human to understand how animals feel when under total biological control. Imagine a human in the process of performing the sexual act and being stimulated during the successive thrusts. The very moment of the pre-ejaculation phase when the human male is totally sexually stimulated, unable to control himself and compelled to proceed to ejaculation, represents the state of mind of the animal when he is about to penetrate the female.

Human Personality in Sex

Human sex is not only the contact of two genital organs but also of two different individuals. Unlike animals, the whole human body as well as the whole personality of each individual also is involved. Everything that makes one individual different from another comes out in sex since, as previously shown, the higher cerebral centers play an important role. Sex becomes an expression of the personality of each individual and, therefore, it is different depending on the various personalities. Assuming animals have a personality, it does not come out in sex.

Sexual Behavior at Different Stages of Modern Life

In humans there are major differences during each life stage because the higher cerebral centers learn to control the biological factors, resulting in conscious and unconscious sexual stimulation at all levels.

Sexual awareness starts very early in prepuberty, long before puberty. Even though the biological push for sex does not appear until puberty when the biological hormones start to rise, awareness of sexual differences and sexual curiosity appears as early as three years of age and is well established at six. During childhood the child has all the time necessary to become familiar with the concept of sex. Sexual consciousness is slowly introduced during childhood. When puberty appears, the young adolescent is prepared and the biological shock is amortized over many years.

There are major differences during adolescence between animals and humans. First, human puberty does not automatically mean sexual activity. Puberty in girls appears between nine and fourteen with an average peak at age eleven. In boys it appears at about the same age or a little later, with an average peak at age twelve. Today, sexual activity is starting in mid-adolescence (between ages fourteen and sixteen), which is becoming a social and psychological problem. Without getting involved in sociological discussions, there is agreement that sex in early adolescence is considered pathological, and sex in mid- and even late adolescence is a concern for everyone. Sexual activity at the end of human adolescence or the beginning of adulthood is considered the norm. The human family unit, as opposed to that of animals, is not disrupted immediately at puberty; parents still take care of their children during and even beyond adolescence. The goal is that sexual impulses should be restrained in early puberty. The adolescent or teenage years are the critical period between biological puberty and full emotional maturity. Biological puberty is the anatomical maturation of the secondary and tertiary sexual characteristics associated with sexual desire and the ability to perform a sexual act. Full emotional maturity or adulthood occurs at the end of the teen period and is characterized not only by a fully matured physical body but also the mental and emotional ability to behave in a responsible manner in the adult world. The relation between sex and adolescence always has been a social problem.

The earlier sexual practice is started in childhood or early adolescence, the more frequently emotional problems occur. No relationship of cause and effect is stated here, whether early sexual practice leads to emotional problems or emotional problems lead to early sexual practice. It is simply stated that they are commonly associated.

A fundamental difference, therefore, is that sexual activity in animals starts or could start immediately at biological puberty. In humans, knowledge of sex is acquired before (but its practice is delayed until after) puberty and during adolescence until further sexual education is acquired and more control of biological sexuality is established.

Sexual education may be very intense in humans especially during adolescence. Adolescents are taught in depth about proper behavior toward the opposite sex. Restraint and control are stressed. Specifically, individualization versus promiscuity, especially for girls, is emphasized. Both mothers and fathers can be involved with educating their sons and daughters. Of course, there are many variations and controversies among humans concerning sex education. The old stereotype of the naive bride and groom is now reconsidered. This is vastly different in animals where, without any sexual education whatsoever, sex is wild very early.

In adulthood, the three factors that regulate animal sexual behavior (biological hormones, physical strength, and hierarchy) also play a role in humans, but they do not have the same dominant role as in animals. Specifically, human sex is available to a majority of individuals, while in animals it is available only to a minority. Other human qualities originating from the higher cerebral centers (intelligence, maturity, emotion) play an important role.

After menopause, human sexuality again differs from the animal one. In humans, sexual practice is usually not abandoned at menopause. This is the most fascinating aspect of human sexuality. In spite of the disappearance of sexual hormones, there is continuing sexual practice. In aging people, health and availability of a sexual partner are more important than age (Abitbol, Davenport, 1974). Sexual activity in the seventies, eighties, and even nineties is a common occurrence. Hormonal factors no longer play a role in postmenopausal human sexual behavior. The biological hormones have disappeared, and the practice of sex is solely under the control of the higher cerebral centers. It is this stage of life that proves that the biological hormones (estrogen and testosterone) play a limited role in the human sexual act, and that the cerebral factors are important. It is ironic that with the administration of sexual hormones in post-menopausal animals, they can resume part of their sexual activity, but the same hormone administration to postmenopausal human males or females usually leads only to emotional disturbances.

Sex in animals is an opportunity to express aggression against females and other males the same way they express it in every aspect of their ordinary life. It is a way to display physical power and superiority. So, the whole picture of animal sex is a negative one, not a positive one.

In humans, it is the opposite. Under abnormal conditions, sex also can produce fear, aggression, submission, and a get-it-over-with feeling, but this is highly unusual. Most of the time love, joy, peace, and general positive feelings are produced. Sex demonstrates how living creatures relate to each other, negatively in animals or potentially positive in humans.

Sexual Taboos and Rituals

A taboo is an interdiction; a ritual is an order in which to perform a specific human behavior. Some are specific to certain cultures, others are general to the whole of mankind. Some have persisted since time immemorial, others have appeared and then disappeared. Taboos do not exist in animals, but some animal species may give the appearance of performing rituals. Human sex is filled with taboos and rituals. The most common sexual taboos that exist today are against sex before or in early puberty, incest, rape, promiscuity, polygamy. The most common sexual rituals are courtship, engagement, marriage, faithfulness, divorce, etc. They are not biological in origin, thus they do not exist in animals. It is obvious that they originate from higher cerebral centers and, again, represent an attempt to regulate sex. Their function is to control the sexual drive, sexual instinct, and the purely biological aspect of sex. They represent, therefore, an interference by the intellectual and psychological over the biological.

Transcendence in Human Sex

In humans, the sexual act becomes the most meaningful event and the most powerful factor determining human behavior, aside from the function of reproduction. It is not only sexuality that has been reorganized, it is the whole way of life that has been reorganized around sexuality. As prehumans and humans progressively detached from the animal life, they reorganized their sexual life and used it like a pattern to reorganize their human life. I will show in the following chapters that human evolution is really human sexual evolution.

The human sexual act transcends the act of copulation. As described by Diane Ackerman (1990) in "A Natural History of the Senses," "Sex is the ultimate intimacy, the ultimate touching when we engulf one another. We play at devouring each other, digesting each other, drinking each other, actually getting under each other's skin ... we open the sealed fortress of our body to our lover. ...During intercourse, a man hides part of himself in a woman, a bit of his body disappears from view, while a woman opens up the internal workings of her body and adds another organ to it, as if it meant to be there all along.

There is therefore, at least potentially, much more in the human sexual act than previously described. The human sexual act can go far beyond the casual aspect of animal coitus. For lack of better words, and to avoid any controversial discussion, it can be said that there is a potential "spiritualization" of the sexual act. It is not always there, but the description of the sexual act would be incomplete if I did not mention a potential transcendence that can carry the human far beyond the present human condition. The sexual attraction that two human beings feel for each other can reach unbelievable dimensions, completely unknown in the animal world and even in the ordinary human world. The attachment between two humans goes beyond the short time during which they are sexually involved each time. In animals, there is no concern beyond the sexual act. One may argue that with such a limited brain, animals are not capable of such an attachment. This is completely wrong because there are other situations where natural animal bonding can easily compare with human bonding, such as mother-infant bonding and a pet (dog, horse, etc.) for humans. Animals are capable of bonding, but they just never express it in their sexuality. Of course, the human sexual act can be solely biological as in animals, but it has the capability, for better or worse, to go far beyond it.

In my opinion, the deep meaning of human sexuality is that it goes beyond the sexual act, intercourse, or ejaculation. These physiological manifestations become the expression of a transcendental relationship, far beyond the involved body contact. As Martin Buber would say it becomes "I and Thou" and not "I and It" any more. The whole potential within mankind can become expressed in sexuality.

The spiritualization of the human sexual act can become a moving force that makes the participants accomplish goals in life far beyond their present capabilities. Sociologists and ethologists have stated that when the sexual relationship becomes a love relationship, it is the greatest moving force in mankind. Human beings can modify their entire behavior for the sake of sex, such as King Soliman the Magnificent of the Ottoman Empire who conquered practically the whole western world and had an attachment for his wife that represents the greatest love in recorded history. Human sexual partners can influence their reciprocal lives, especially when there is continuity, to a great degree. It is my opinion that this sexual transcendency is one of the driving forces that made human evolution possible. This sexual evolution is still not complete, and each human being is continuing his or her search, sexually speaking.

Human sexuality and animal sexuality are completely different, and one could not have been derived from the other. The truth is that animal sexuality had to be completely erased for human sexuality to be built on an entirely new foundation. The two systems have absolutely nothing in common. However, although the animal system is altogether restrained and contained in humans, one should never forget that it is only skin deep. It does not take much for humans to revert to the animal system. Human sexuality emerged against a background of animal sexuality the way light emerges against a background of darkness. Like all great accomplishments, human sexuality is fragile, unstable, easily disorganized and lost, and the primitive chaos of animal sexuality is easy to return to if not properly contained.

In my opinion, the unfortunate inability to fully reach this spiritual dimension explains the constant sexual frustrations of present humans, so they revert to animal patterns and praise and prefer them. It is sometimes difficult to find the middle ground, and the tendency is to fall within extremes.

Summary

In summary, human sex is diminished biologically, increased psychologically. It represents a struggle between biological and psychological, an attempt to individualize sex, and a desperate and often failed attempt to transcend the biological aspect. Indeed human sexuality is a problem, and indeed animal sexuality not being a problem may appear attractive (de Saint-Exupery, 1979).

Introduction

In animals, hormonal levels (testosterone in males and estrogen in females) in the circulating blood are nil outside the mating period and very high during sexual activity. In humans, they are constant in men and cyclical in women, but they are not higher during sexual activity and lower outside of it. Sexual activity is, therefore, associated with high hormonal levels in animals, not in humans. Specifically, sexual climax in humans is not associated with a high level of hormonal activity (Ackerman, 1990).

Obviously a tabula rasa has been made of the old estrus system of perineal congestion and pheromone production and a completely new system to secure female stimulation and male response had to be built ex vacuo. Human females have accepted the fact that the powerful effect of pheromones on human males is extinct, but some are still looking for a "magic" scent (Ackerman, 1990).

The following table summarizes the sexual differences between animals and humans. Although there are exceptions, they are rare and doubtful in my opinion. Most of the confusion arises when human attributes of sexuality are applied freely to animal sexuality.

	Animals	Humans
Estrus	+++	+or absent
Copulation	Rapid, brutal	Slow, gentle
Pleasurable Sex	---	+++
Knowledge of Reproduction	No	Yes
Involved Senses	Taste, smell	Sight/touch/hearing
Brain	Midbrain	Cortical brain
Sexual Stimulation	Unilateral (fem. to m.)	Bilateral
Transcendence	Nil	Possible

Captive animals are constantly fed and even overfed. This may be another reason why captive and domestic animals do not have a seasonal estrus and are in constant estrus. Higher primates also menstruate cyclically like humans, although as in other animals, it may be seasonal for some. It is not certain whether this human-like menstrual cycle exists only in captivity or in semi-captivity.

The constant misinterpretation of animal sexuality is what I call "reverse" evolution. All differences between the two sexualities are human acquisitions and therefore do not exist in animals. Yet, there is hardly a single aspect of human sexuality which has not been projected onto animals. I give as an example sexual orgasm, a complex behavioral pattern specific onlyh to humans. Another misnomer is courtship in animals.

The simple basic observations reported here sometimes appear to be ignored even in scientific circles. An article in the New York Post (6/25/92) reported that Japanese scientists were showing XXX-rated video films to zoo gorillas to stimulate their low sexual interests. Of course, there were no positive results! It is surprising that no one thought of isolating pheromones from female gorillas in heat and making the male gorilla constantly sniff them.

A serious error to avoid when studying animal sexuality is therefore to look at it in human dimensions. Human sexuality can consecrate the relationship between two individuals of the opposite sex. Any objective observer who expected this human dimension to be present or reported in animals would be disappointed or fantasizing. Animal sexuality will be well understood only when one stops looking at it through human eyes, when one stops considering animal sex in the light of human sexual experience, and when one stops desperately searching in animal sexuality for what may be lacking in human sexuality. If anything, animal sexuality will teach us the regressive pattern of human sexuality, not the progressive pattern. Comparing animal and human sexualities is like comparing the different noises made by animals to human language. Animal sexuality is where human sexuality is coming from and not where it is going. Our sexuality is a human acquisition, such as erect posture, speech, laughing, crying, thinking, and concepts of beauty and art, space and time, etc. Animals do not have sex, they just have coitus; they do not relate to each other, they just copulate.

Additional Characteristics of Human Sexuality

Of the many sexual differences between animals and humans previously enumerated (the list is certainly incomplete), human sex has acquired major traits, of which there are no traces in animals.

First, there is an elaboration of the human sexual act. It is a misconception to think that any excessive physical activity during the sexual act is more animal than human. It is wrong to think that people who spend time engaging in any stage of the sexual act or in performing sex in general, are animalistic in their sexual behavior. It is wrong to imagine that people who gesticulate excessively and scream loudly during sex are behaving like animals. The rudimentary animal sexual act is short, silent, casual, and associated with no motion except for a few successive thrusts. In humans it is a very elaborate physical and mechanical sexual act. Most important is that human coitus lasts longer and has greater variety. Any excess in its duration should not be considered an animalistic regression but rather a superhuman and ultrahuman reaction. This does not mean that overindulgence in the physical aspect of sex is very human; it would be considered a pathologic excess rather than a pathologic regression to an animal level.

A specific point also should be clarified here. A lion or a ram during the mating season can have coitus up to 100 times per week, thus giving the impression of overabundance of physical sex. The other side of the story is that each coitus lasts only a few seconds totaling a few minutes a week, so that a human being having weekly coitus lasting 10 minutes is having more sex, timewise, than a lion or a ram during the mating season.

Second, constant preoccupation with sex modifies and invades the whole way of life of the human sexual partners. The animal can be occupied but never preoccupied with sex. With humans it is the opposite. Humans are preoccupied with sex well beyond the time devoted to sexual intercourse and their sexual needs. Women appear to be more preoccupied with sexual attraction and men more preoccupied with sexual performances, but this is not a general rule.

Finally, because they can control their sexual behavior, humans, not animals, are responsible and accountable for their sexual and any other behaviors for that matter. This notion of responsibility is fundamental for understanding the problems of human sexuality. Nature handed us a simple system of sexual behavior: a biological (animal) one made of seasonal stimulations, biological calls in the form of estrus, a scent to recognize and locate the sexual partner, a pattern of sexual behavior rigidly predetermined, and last but not least, one with no commitment of any kind. With this system there are no decisions to make, no responsibility to take, no questions to be asked, and one can be completely passive and just follow the natural process. I do not know how much animals are conscious of their sexual behavior or any other behavior. It does not matter because they cannot control it.

Humans, on the other hand, are conscious of their behavior, can control it, and are, therefore, responsible for it. In humans, with the collapse of the whole biological process, there is no mating season, no biological need for sex because of weakened estrus, no scent indicating female receptivity or male readiness, and no fixed pattern for sexual behavior. Now the human sexual partners need to search for each other sexually and otherwise. The problem becomes more complex because important questions arise: when to have sex, where, how, with whom, and, worst of all, why. I do not consider procreation enough motivation for sexual activity. Having discussed this problem with thousands of patients, I can declare without hesitation that sexual activity is usually not motivated by the desire to get pregnant. Since the biological call is not enough motivation or explanation, humans have to build their own sexual system on remnants of the biological one. It is difficult to ascend to a new sexual system without a struggle against the existing one. To quote Antoine de Saint Exupery (1979), "There is no ascension without fighting against gravity." No wonder animal sexuality is so attractive to some humans! To describe human sexuality one must include this struggle and the desperate and often unsuccessful attempt to substitute a new system for the biological one.

The Basis for Sexual Evolution

How animal sexuality evolved into human sexuality is a fascinating subject because a radical transformation actually occurred. Now that the two sexualities have been compared and differentiated, the sexual evolution from the first to the second type can be approached.

It is incredible how difficult the study of sexuality can be. Let us remember the controversy that surrounded Darwin in the beginning of his work on evolution, a concept readily accepted today. When it comes to sensitive subjects such as behavior or sexuality, the concept of evolution is met with similar resistance. While the concept of anatomical evolution is accepted, behavioral evolution is not readily accepted. From an emotional point of view the subject of sexuality is a highly charged one, and today the average person is completely bewildered as to what human sexuality is or should be. The present study is a difficult and sensitive one from intellectual and emotional points of view, and it is done with the understanding that there will continuously be controversy, challenge, and criticism.

The only common thread between animal and human sexuality is that vaguely similar sexual organs are involved. Should the term sexuality be used for animals and humans just because the same organs are used? Animals and humans have tongues and lips; animals use their tongue and lips only for tasting and eating while humans, although they keep the tasting and eating function, use them mostly for speech. Animals use their gustatory and olfactory functions chiefly as a mode of communication and relationship while these functions have become dispensable and esthetic ones in humans. Therefore, it would not be surprising if sexuality, exclusively used by animals for reproduction, is also used by humans for yet other purposes. How this brutal male beast and terrified female have (or have not) been completely transformed into actors of modern sexuality is indeed puzzling. The various and successive steps of this transformation or metamorphosis are difficult to unravel. Let us attempt anyway.

How this transformation occurred obviously has to do with how animals differ from humans. This statement simply means that as animals became transformed into humans, animal sexuality became transformed into human sexuality. Therefore, it is essential to know the main anatomical differences between animals and humans to determine how they affect the sexual differences. This list could be endless unless limited to the basic traits influencing the transformation of sexuality. As discussed in the first chapter, the two major characteristics of human evolution are bipedal behavior and encephalization (brain hypertrophy) which played an essential role in the transformation from the animal to the human condition, and, therefore, also play a role in transforming animal sexuality into human sexuality.

Bipedal behavior tremendously influenced the pattern of sexuality in two ways. First and most obvious was reconsideration of the sexual posture. Animals perform the sexual act in the ventral-dorsal position (mounting). This general rule is a consequence of quadrupedalism. When hominids stood with straightened bodies, sexual posture had to be reevaluated. This will be investigated in detail in the next chapter.

Erect posture also influenced sexuality in the hominids with the appearance of new organs: the hands. Nonhuman primates use their four limbs for posture/locomotion as well as manipulation. The hominids use the hind limbs exclusively for posture/locomotion and the front limbs for manipulation. Exclusive bimanual behavior is just as characteristic of hominids as is bipedal behavior. The hands are perfect for manipulation the way the feet are perfect for posture/locomotion. Touching in the human species is done exclusively by the palm of the hands which has become perfected in this function. The resulting introduction of manipulation by hands and touching plays an important role in the evolution of human sexuality.

Encephalization is the other characteristic of hominid evolution. During evolution, the brain matured through superimposition of layers (Armstrong and Falk, 1982).

Present humans are called "*Homo sapiens sapiens*" (HSS), which means they have a superior intelligence. Without getting involved in present day discussions about the left half of our brain being reserved for intelligence and the right half for feelings and emotions, it is obvious that our feelings and emotions have progressed the same as our intelligence (De Unamino, 1954). Today, some people argue about whether we really have progressed in terms of feeling and emotions compared with animals, and some animal lovers deny any progression and even speak in terms of regression. To avoid arguments let us agree that our emotions are different and they have, therefore, evolved. There has been an evolution (if not progression) in the two sides of our brain, intelligence on one side and feelings and emotions on the other side, when hominids evolved from an animal-like condition to the present human condition. Our superior intelligence and superior (or just different) feelings and emotions reflect on general human behavior and consequently will have a marked effect on human sexuality. These characteristics (erect posture, use of hands, superior intelligence, and different emotions) are the basis for sexual evolution.

Chronological and Logical Approaches

We can assume that each step in the anatomical transition from the animal to the human condition was associated with a corresponding transition in sexuality. It would be fascinating to reconstruct the step-by-step transformation from one type of sexuality to the other, the emergence of the new sexuality, and the abandonment of the animal one. At the present time, a direct study of such a chronological sequence is almost impossible and a different approach has to be devised. In this study, the successive events will be reconstructed by completely disregarding a chronological schedule and following what I call a logical sequence. By that I mean an attempt will be made to put together facts that have some similarity although they may have been very distant timewise.

Sexual touching is an example. A chronological approach would attempt to describe when the acquisition of various steps of sexual touching occurred. The logical sequence does not tell who was doing what, at what time, during the hominid evolution because it is hard to determine if the changes occurred abruptly or progressively, together or separately. In the chronological sequence it is, in fact, difficult to determine which occurred first, next, and last. Therefore, to simplify this study, these changes will be presented in a logical sequence. Hopefully a logical sequence will allow better analysis of each step, permitting the reader to determine the factors surrounding each change without being burdened by the chronological determination of each step. At the end of this analytical study, an attempt will then be made to place the different steps chronologically.

The closest animals to *Homo sapiens* are the higher primates (chimpanzees, gorillas, and orangutans). Among these three species, the chimpanzee is considered to be the closest. Based on a starting point with a creature similar to the present chimpanzee, the two events, first, erect posture and second, brain enlargement, appeared in that order and have been more or less accurately described by anthropologists. The gradual formation of the anatomical appearance of the present *Homo sapiens* has been more or less reconstructed step-by-step after the discoveries of different fossils in Africa, Asia, and Europe. Based on this anatomical evolution, an attempt will be made to reconstruct the sexual evolution.

Rationale for the Study of Sexual Evolution

How general behavior evolved from a chimpanzee-like creature to present *Homo sapiens* is still a mystery. The discovery of fossils helps to reconstruct the bony structure and even the whole body of the intermediate species, but reconstruction of the evolution of the mind, of thinking and feeling, and the gradual appearance of these human traits is speculative. Based on meager data, some investigators have been able to reconstruct aspects of life of various extinct hominids, but the evolution of general behavior of the hominids, the gradual appearance of our present feeling and thinking, which could be called the birth of human spirit and consciousness, has been out of the reach of researchers. Such failure may have to do with the concept of research. Anatomical scientists refuse to advance any theory or speculation on these subjects unless based on rigorous scientific data (archeological findings). Archeological findings of extinct hominids of the last three million years consist of a pile of bone fragments all of which could be easily placed on a table; there is very little to reconstruct in the way of life of extinct hominids. However, the point of departure (present chimpanzees) and the point of arrival (present *Homo sapiens*) are both accessible and easy to describe. To fill in the gaps, a tremendous amount of speculation and imagination is necessary.

The following description of the birth of human sexuality may seem like folklore. An attempt to base the following descriptions on as much data as possible, to be as rational as possible, and to abstain from any haphazard speculation, is my goal. There is nothing wrong in speculating, however. As mentioned by Charles Darwin (in a letter to Alfred Russell Wallace): "Without speculation, there is no original observation." Quoting Oscar Wilde: "There is no geographical map without utopia in it."

Is human sexuality derived from animal sexuality to which new human elements have been added, or is it so radically different that animal sexuality first had to be completely eradicated to build the new sexuality? If the first concept is correct, the title of this book should be "The Evolution of Human Sexuality." If the second concept is correct, the title should be "The Birth (or the Emergence) of Human Sexuality." It will be shown later that the second concept is correct.

Coitus is the intromission of a male penis in a state of erection into the vagina of a female, with the sexual act being terminated after ejaculation of sperm into the vaginal cavity. Mechanically, therefore, this part of human sex is similar to that of animals; the similarity stops right there however. As a marriage counselor, I frequently heard a typical story of a woman frustrated when the sexual act was terminated by sperm ejaculation. Yet, when I observe an animal sexual act on videotapes, in the zoo or in jungle habitats, I have never seen animal coitus lasting beyond ejaculation or a female animal expressing, even remotely, anything that could resemble sexual frustration.

One characteristic of primate societies, specifically the chimpanzee society, is that they are great imitators. Their curiosity always is intense and whatever one individual does, another individual wants to imitate, so that any innovation of behavior in one individual immediately finds candidates among others. This curiosity is general, but it is easy to imagine that it is more intense when it comes to sexual behavior, and any sexual innovation by an individual will be repeated, or at least attempted, by others.

I will envision a male hominid, named Henry, traveling through the successive steps of hominid evolution. He will be traveling with his female companion, Henrietta, also going through her own hominid evolution. During a three-million-year voyage, Henry and Henrietta underwent significant physical changes to their bodies and behavior (sexual or otherwise). Thousands and thousands of times, Henry and Henrietta died and were reborn, each time undergoing abrupt or gradual subtle minor and major changes. In the beginning, Henry and Henrietta were like the present chimpanzee or "Lucy," and at the end of their voyage they were like the present *Homo sapiens*.

ANTERIORIZATION AND ITS CONSEQUENCES

New Sexual Posture

The interesting aspect in the evolution of sexual posture from ventral-dorsal (mounting) in animals to ventral-ventral in humans is that there are no transitional or in-between phases. This is different from the acquisition of erect posture, for instance, where all the intermediary stages between quadrupedalism and bipedalism can be described. To practice ventral-ventral posture, ventral-dorsal posture has to be eliminated. It is one or the other, there is no intermediary sexual posture, although some early hominids may have practiced both sexual postures interchangeably.

Today, monkeys, apes, and all nonhuman mammals in general have a ventral-dorsal coitus, although some exceptions on either side could be cited. The reasons why animals practice ventral-dorsal coitus is easy to explain. It is more natural since the female does not have to move from her natural quadrupedal posture. In fact, some animal females go on with their usual activities, such as eating, browsing, grooming, etc., while being mounted and the male just jumps on her back and takes her by surprise, thus securing her cooperation more easily. During mounting, the female, who is usually weaker, exposes only her back consisting of the vertebral spine, the occipital part of the skull, and the strong muscles of the back of the neck and the thorax. A bite by the male can do only limited harm to the thick skin. In ventral-ventral sex, animals would expose the face, eyes, snout, front of the neck with the carotid and jugular vessels, trachea, and most of all the belly; this could be very dangerous. A strong bite on any of these organs could produce severe damage and even be fatal. Therefore, the animal female is safer when exposing only her back. Besides, why face-to-face? What would be the advantage? Sex in animals is a scent affair, and it is simpler to sniff posteriorly than anteriorly. Sex in animals is not visual. What the rest of the female body looks like is of no interest whatsoever to the male and the female is only interested in taking a protective attitude. The purpose of ventral-ventral (face-to-face) sex is to have visual identification and auditory communication, as will be described later.

Consequences of New Posture

When Henrietta stood up, coitus or penetration of the penis from the back of the female became more and more difficult. Henrietta and Henry, her male sexual partner, had their first sexual problem, but unfortunately not the last one from there on. For the first time in animal history, the sexual posture had to be reconsidered. Several hypotheses can be evaluated. Henrietta could have reverted to quadrupedal, hands-feet, or hands-knees posture for coitus, or taken a lateral posture with a ventral-dorsal coitus, or Henrietta and her male partner could have taken the great leap and adopted ventral-ventral sex.

There are arguments for and against each of these theories. Against the ventral-ventral posture is the fact that it requires the female to lie on her back. This posture is sometimes seen in gorillas (Fossey, 1983), rarely as a sexual posture in chimpanzees except as part of playing and rolling. Among living primates, sleeping in supine posture is seen only in humans and even then not as a rule (Abitbol, 1989). To practice supine posture, the whole trunk has to be wider laterally than anteroposteriorly. When Henrietta ("Lucy") was at the beginning of her journey, her pelvis was very wide laterally and very narrow anteroposteriorly, but her chest was wider (or at least as wide) anterioposteriorly as laterally (Schmid, 1983). Therefore, it is hard to determine if Henrietta could have easily laid on her back and performed sex in this posture. It is only with the next step in hominid evolution *(Homo erectus)* that these hominids could have taken the supine posture.

Humans sometimes practice ventral-dorsal sex. If the woman lies prone (ventrally), intromission of the penis in the vagina is difficult in such a horizontal posture. For ventral-dorsal sex to be possible, the female has to take a knee-chest posture and spread her knees. The female has to carry her head forward and her arms are not accustomed to carrying so much weight.

It is not important when ventral-ventral sex occurred during the hominid evolution or whether it occurred abruptly or progressively. The movie version of "Quest of Fire" beautifully and romantically described the change from ventral-dorsal to ventral-ventral sex as taking place during the Neandertal stage, specifically at the request of the female. The crucial event is what we will call anteriorization (ventral-ventral) of the sexual act as opposed to the animal (ventral-dorsal) coitus. The female now lies on her back and the male lies with the front of his body over the front of her body. Once this anteriorization has been fully accomplished, it has multiple consequences.

The first and fundamental consequence is that the female is freed from bearing the whole weight of the male. Although part of the male body is on her chest and pelvis during ventral-ventral posture, she does not have to carry the entire male body and contract her hind limb muscles. This is a fundamental point because the female no longer concentrates on carrying the whole male body, and ending the sexual act quickly; she is free to concentrate on whatever she wishes.

The second fundamental consequence is that for penis intromission and successive thrusts to be possible, the female has to deliberately spread her thighs and legs in complete abduction. If she refuses, penetration will be difficult, incomplete, and even impossible. Therefore, the female has the potential of accepting or refusing the sexual act.

The third fundamental consequence is that the male partner is also free to concentrate on whatever he wishes since he does not have to force the sexual act on the female.

The fourth fundamental consequence and, in my judgment, the most important, is that the female now sees the sexual partner and what is happening during the sexual act. The sexual act now becomes visual. For the first time in the history of sex, the two partners can see and identify each other.

The fifth consequence is additional body contact. True, there is some body contact in animals but it is directed toward immobilization. The sexual act is now tactile, involving a large segment of the body, at least more than in animals.

The sixth and last consequence is the freeing of hands. It reinforces body contact by allowing reciprocal exploration of bodies. It allows embracing and intimacy or, conversely, it allows rejection. Body contact in human sex occurs when the two sexual partners are naked as well as when wearing some clothing, although the contact feeling is more intense with nudity.

The animal female does not have a posture specifically reserved for coitus. In humans, it is completely different. During sex, the female lies on her back with her lower limbs partly flexed and apart. This is a posture the human female takes *only* during the sexual act and at no other time in her life, making the posture highly specialized for the performance of the sexual act. So, the sexual act, from a casual one has become a specific one. In this chapter, I will describe the intellectual aspect of anteriorization. In Chapter XII, I will describe the emotional aspect.

Touching in New Posture

In nonhuman primates, the skin is too thick and too hairy for any refined physical contact relationship. Natural selection for hair only on specific parts of the body can be speculated on. For women, more than men, selection was for a non-hairy face. Sexual stimulation using touching is born, replacing pheromones. The expansion of touching to the level it has reached in the present *Homo sapiens* is sexual in origin and started with the ventral-ventral sexual act. The hominids expanded it to all aspects of their life and did not limit it exclusively to sex. Touching, in fact, was the most important substitute for pheromone stimulation, which was disappearing.

Disappearance of Aggression

In animals, sexual activity is rather external occurring somehow outside the body. In humans, the sexual act occurs inside the female. It becomes an internal activity, with the most intimate aspect of her body fully involved. In ventral-ventral sex, the male still may keep the illusion of dominance because he is often on top, but he has actually lost control because he cannot force the female to have sex and now needs her cooperation. Therefore, aggression had to disappear. Remember, the sexual act in animals is aggressive, painful, and rape-like in nature. If we transferred this brutal behavior in face-to-face sex, it could be dangerous and no longer possible. Selection was not in favor of this type of male aggression and aggressive males would have to disappear in favor of gentler males, at least potentially.

It is not known whether the size of the larger canine teeth in relation to other teeth disappeared with Henrietta or with any other known form that preceded Henrietta. With smaller canine teeth, male aggression was not as dangerous and in battles between male and female, the canine teeth lost their major role. There still was biting but it was not as deadly and females could stand up to a fight, making rape more difficult.

Origin of Language

At the origin of the hominid evolution, just prior to and even in the beginning of Henrietta's era, the hominids did not possess any obvious language. The intimacy of the sexual act in the new ventral-ventral posture must have given a special meaning to any noise or vocalization made by either of the sexual partners. In general, sexual relationships required other modes of communication (how to signal sexual attraction and rejection, readiness for the sexual act, emergence of climax, etc.), especially after the disappearance of pheromones. With depheromonization (see later), any new mode of communication would do, especially since female cooperation was more and more necessary. Hearing, listening, and vocalizing started taking on a special meaning before and during the sexual act, and easily would spread to other aspects of daily life. It is not stated here that sex is at the origin of speech and hearing but, during a sexual relationship, speech and hearing were more freely used as a mode of sexual communication than with the silent mounting position. This could have been only one of the factors that initiated the appearance of language.

Identification of Sexual Partner

The animal female barely knew who mounted her and every mounting was alike. As far as the male was concerned, choice was determined more by proximity, availability, fighting, etc., than being directed to a specific female. With face to face posture, identification of one sexual partner, different from any other individual of the same species, becomes possible. Recognition of an individual who is different from all others is a crucial step in hominid evolution, and most likely is sexual in origin.

Depheromonization

At the beginning of, or shortly before, the hominid evolution, sexual attraction for Henrietta or Henry was similar to what we currently see in higher primates such as the chimpanzee, gorilla, or orangutan. The early Henrietta in estrus goes through a perigenital congestion and produces pheromones to which the early Henry was attracted. The olfactory attraction to pheromones is directed toward the perigenital area and not the rest of Henrietta's body (face or otherwise). With this sexual pattern in mind, let us see the changes that occurred as a result of Henrietta standing up and the anteriorization of the sexual act. When Henrietta stood up, Henry had to bend down to sniff Henrietta's perigenital area. Henry is used to Henrietta's quadrupedal posture, which facilitates sniffing and mounting, and he must have expressed his confusion by becoming very aggressive. The solution to the confusion and aggression is depheromonization or abandonment of pheromones as a mode of sexual attraction. Eventually, other modes of sexual attraction are to appear, and it is not an olfactory perception of external genitals on which the female hominid will base her sexual attraction. The female hominid wants to be identified individually. Pheromone production and direct attraction to external genitalia are substituted with visual attraction (face, eyes, rest of body) and tactile stimulation.

The eyes play a fundamental role in replacing pheromones. Through the eyes, a woman can signal either sexual attraction, repulsion, or no interest. The variety of messages humans can send through their eyes is tremendous. Pheromones produce general and total attraction to all males without selectivity, while human eyes can indicate selectivity. Pheromone production quickly had to be replaced by sexual attraction of visual nature, which Henrietta could control better.

Modern women have lost pheromones, which used to put the male in a state of frenzy. Pheromones have four disadvantages. First, production is a biological phenomenon independent of the female's control. Second, it is produced in the perigenital area and modern women would prefer another area of their anatomy to be sexually attractive. Third, male reaction to pheromones is too excessive and leads to violence. Fourth, it acts on all males and cannot be directed to a specific individual. Pheromones are out, although more moderate agents, such as perfumes, will be used by modern women.

During the hominid evolution there was an almost complete collapse of the hormonal system that regulated sex in the animal female. Female sexuality had to be totally rebuilt on an entirely different basis; namely, giving up pheromones and introducing vision, touch and sound.

Women want to keep themselves sexually attractive, but without pheromones. An olfactory signal revealing their inner sexuality would be completely out of the question. Nothing would offend a woman more than an external appearance similar to the effect of estrus in animals.

The problem for Henrietta when she stood up was to substitute another form of stimulation, one more refined than the dangerous pheromones. The search for this magic weapon started when Henrietta and Henry switched to front-to-front sexual posture and has been ongoing since.

Animals have a sense of smell that is a million times more powerful than in humans. It is a relic in humans. All relics become valuable; their meaning is fantasized, idealized, and even distorted. That the animal's sense of smell consists mostly of sniffing each other's stool and urine (essentially each other's genital and anal area) and for hunting and killing is all but forgotten.

What has been said about smell also applies to taste. It also is a relic that has partially disappeared.

How would modern women react if men approached them not by looking at their beautiful face or listening to their musical voice, but by bending over and sniffing their anus and vagina! My opinion is that it is just as well we lost most of our sense of smell and a great part of our sense of taste. If we had retained them as they were, they would have distracted us from expanding vision, hearing, speech. In brain anatomy, expansion of the latter was possible only with a regression of the former.

Sophia Grosjman, a fragrance specialist, was interviewed by Diane Ackerman. She said, "I would like to make a perfume some day so seductive to man that no woman could be resisted. It would be the most incredible thing I could do in my life. This is not a professional feeling it is strictly a female feeling." What Sophia forgot is that humans have lost most of the brain zone connected with olfaction and what she is talking about is return to the animal condition.

Ackerman (1990) calls the sense of smell "the fallen angel," "a relic," because over time it has become the least necessary of our senses. Quoting Ackerman, "evolution has phased out our sense of smell, chemists have labored to restore it...we just cannot let go of it...we will not be weaned."

In kissing, we taste, smell, and touch each other all at the same time. When associated with plunging of the tongue, it is a form of sexual intercourse (Diane Ackerman, 1990). Kissing is practiced all over the world but only in humans; what is reported in animals is actually sniffing and licking.

Professional smellers (or smell detectives of bad smells; Modern Maturity, Feb-March 1993) are just as common as professional perfume smellers; they just are not as popular. Westerners find fecal smells offensive but Masai (Ackerman, 1990) like to dress their hair with cow dung which gives it an orange-brown glow and a powerful odor. Children like most smells until they are old enough to be taught differently. In Western society, passing gas is considered repulsive and discourteous. It is important to note that repulsion is mostly for another's fecal smell since each is very tolerant of his/her own.

Deseasonalization

Once pheromones disappeared, an important revolution fundamental in defining human sexuality occurred. Seasonal sex disappeared and was replaced by year-round sexual activity. Women are always women, and men are always men. Sexual attraction can be activated before puberty and persist well beyond menopause, since it is no longer hormonal, but visual, tactile, and so on. I cannot stress strongly enough that the radical difference in sexuality between animals and humans is that in animals male and female awareness occurs after puberty and is only seasonal, while in humans the male and female awareness continuously occurs before puberty, all year round during maturity, and continues well after menopause. The consequence is sexual awareness of the human population almost from early childhood to death.

THE NEW PERIGENITAL AREA

Introduction

This chapter discusses the anatomical proximity of the vagina to the anus and urethra. It also discusses the problems arising from the physiological proximity of urinating and defecating to the organs of sexuality. One may wonder what my reasons are for putting these functions together. I understand that such a subject is embarrassing for most people and repulsive to some. I, therefore, suggest that those readers who may feel uncomfortable with such a discussion, just skip this chapter and go to the next one.

"On the Origin of the Species" written by Charles Darwin in 1859 represents one of the greatest scientific achievements in the history of mankind. Charles Darwin and his contemporary, Henry Wallace, shocked not only the scientific community but the whole world. The repercussions are still felt today. Before Darwin's time and thanks to the chapter on Genesis in the Holy Bible, men and women were proud of their bodies and minds which they considered to be a creation and a gift of God in all their minute details and completely different from all the beasts which God created in bulk in one day. A spark of God was placed by God Himself in this sublime creature which was carefully planned in a specific time, placed in a specific location (Garden of Eden) for specifically designed purposes known only to God. Now Darwin said that we descended from the apes. Cartoons joked about Darwin's discoveries and some showed a society of apes complaining that humans could represent their progeny (Leaky, 1986). This was nothing more than fear and rejection of apes being our ancestors. The most significant statement on our descendance from apes was made by the wife of the Bishop of Worcester who said: "Descended from the apes! Let us hope that it is not true, but if it is, let us pray that it will not become generally known." What was it that this lady was afraid of that might be known? Possibly she was concerned that all the multiple layers of clothing she was wearing from head to toe were just hiding some animal ugliness underneath, and that all the fancy clothing the gentlemen of that time wore (including her husband, the Bishop of Worcester) were *covering* some beastly anatomy which was not to be exhibited. Suddenly it became obvious that men and women were wearing large and modern fig leaves to cover their shame. Suddenly it became obvious that the reason humans practice sex in privacy and darkness is in order not to be aware of similarity between human and animal sex. Suddenly the ugliness of the most intimate parts of the human body became obvious, to quote Camille Paglia (1990). Suddenly the human body was no longer perfect, not a perfect divine creation but an ugly and imperfect remnant of animal nature, something as dysfunctional as an animal body. One needs time to recover from the traumatic realization that we are not divine, but animal, in origin.

The most fascinating aspect of evolution is its evolving concept, i.e., things do not have to be what they are and since humans are not perfect creations of God, then one can criticize *Homo sapiens* in all respects without the fear of being burned alive as a heretic. Men and women lost their divine ancestry and substituted an animal ancestry. The reason for the fascination of our generation for the animal kingdom now becomes clear.

The "Bizarre" Location of the Human Genital Organs

In this chapter, I will comment on men and women's unjustified feeling of "shame" and their reaction to the anatomical proximity between the organs carrying the functions of urination, defecation, and sexuality. This specific anatomy markedly contributed to the formation of human personality, as reported by Freud (1963).

If not aware of our evolutionary history, any clear thinking and intelligent person would find the location of the vagina (between urethra and anus) puzzling and would wonder about the copulation function located between and in close proximity to the functions of urination and defecation (Ackerman, 1990). On the male side, the the penis is located somewhat distant from the anal orifice, but his combined urination-ejaculation in the same organ is puzzling. The classic explanation, that the specific location of the genital organs and functions is a remnant of the evolutionary process, barely relieves the embarrassment and anxiety.

There is little difference and even some similarities in the way animals copulate, urinate, and defecate. Also for these organs to function properly it is necessary they be anatomically close in animals since pheromones are present in urine, feces, and vaginal secretions. Since pheromone-producing glands are located in the perianal as well as perivaginal areas, urine and fecal smells most likely are part of the total olfactory component attracting the animal male to the animal female genital area. Therefore, the location of the vagina between anus and urethra in animals is logical and necessary. In humans, pheromones have disappeared, fecal and even urinary odors have become obnoxious, and direct sexual attraction to the current female genital area becomes a problem. Humans have inherited an anatomy that no longer corresponds to the present human sexual conditions. I sometimes wonder how much our sexual psychology or, rather, our whole way of life, would have been changed if our genital organs were located distant from the organs of urination-defecation.

Humans are now 'stuck' with this specific anatomy, with the vaginal orifice located between anus and urethra, with the function of copulation performed between the organs of urination and defecation, and with a penis carrying the urinary function. The truth is that the consequences of this complex anatomy are far reaching in humans.

"We are born between urine and feces." This statement was attributed to St. Augustine by Sigmund Freud. St. Augustine who, prior to his conversion, was notorious for his libertine way of life could just as well have said, "We perform sex between urination and defecation" (St. Augustine, 1960).

Animals neither hide nor expose any part of their body. Tthey will indifferently sniff and/or lick each other over any part of the body, from head to uroanogenital area. In humans, certain zones, specifically the uroanogenital area, are taboo and to be shielded completely from any exposure or touch. The three functions, defecation, urination, and the sexual act, are performed privately, while animals indifferently expose all these functions.

Regardless of when these taboos appeared during the hominid evolution, they are still very much alive today and remain essentially intact, although with many cultural variations. Visual exposure of defecation and urination is much more restricted in the human female than the human male who seems to be more casual about it. Here too, there are tremendous cultural variations. When I show a female patient an anatomical diagram displaying her genital organs in detail and the relationship to anus and urethra, there is a lack of interest or a shyness, with no desire to go any further. Modern teaching in elementary and high schools is more liberal today and includes studies of anatomy and physiology of the genital organs. My personal observation is that there is more shyness displayed by girls than by boys who attend these courses. How, when, and why taboos in exposing these different organs and functions appeared needs to be investigated.

Anatomy

In the human female the three orifices (urethra, vagina, anus) are in very close proximity and may lead to anatomical confusion. Women themselves often have difficulty in distinguishing the different orifices. As a gynecologist, for instance, it is not unusual for me when treating a patient for abnormal bleeding to find this woman unable to tell me if the blood is coming from the urethra, anus, or vagina. The human female often has difficulty recognizing that the urethra through which urine is passed is located anteriorly, the anus through which feces are evacuated is located posteriorly, and the vagina through which the menstrual blood flows (and sexual intercourse is practiced and fetus is expelled at the time of delivery) is located in the middle, between the urethra and anus. I have heard a few teenagers describe their first sexual experience explaining that they did not know "where it went," neither juvenile sexual partners knowing exactly where the erect penis was supposed to penetrate.

The male anatomy also can lead to confusion, although to a lesser degree. For instance, a man will never confuse bloody stools with bloody urine, but he might be unable to distinguish bloody sperm from bloody urine.

Function and Posture

The problem becomes even more delicate when one considers the posture taken when performing the function of these three organs. In this matter, comparison with nonhuman mammals and evolutionary considerations become necessary (Table 1). I will stress again and again that the major difference between men and women is that men urinate anteriorly, while women urinate posteriorly; both defecate posteriorly; both copulate anteriorly with the specification that, compared to animals, women anteriorized copulation while in men it always has been anteriorly.

Table 1. Comparative Function in Animals and Humans

(A=anterior function, P=posterior function, N=neutral between anterior and posterior)

	FEMALE		MALE	
Function	Animal	Human	Animal	Human
Defecation	P	P	P	P
Copulation	P	A *	A	A
Urination	N	P+ *	N	A

(Comments on Table 1. The human female became similar to the male when she "anteriorized" sex, but maintained the difference when she definitely "posteriorized" urination. The human male did not change much from the animal condition. Posterior defecation remained similar for all.)

Sexually, the animal female is approached posteriorly (P), but the animal male coitus is definitely an anterior function (A), since the erect penis takes a position located anteriorly to the hind limbs. Most animals, males and females, urinate by lifting one hind limb or by spreading the hind limbs and urinating between the two hind limbs. Urination in quadrupeds is neither anterior nor posterior (neutral or N). These anatomical considerations are not important to animals who barely have a direct vision of their genital organs.

Erect posture revolutionized these functions more in the human female than in the male. In the human female, coitus is now an anterior act (A) similar to the human male and different from the animal female. As far as urination is concerned its posterior (P) aspect is more marked in the human female compared with the animal female since there has been an increased posteriorization in the posture taken by the human female during urination. The reason is very simple. If a woman urinated standing up like a man, urine would run down her legs. Women solved this problem by crouching or sitting on a bowl when urinating, thus taking a position similar to that of defecation (P).

In the human male, there has been less change in postures from the animal condition. Coitus is still anterior (A), and urination is definitely anterior (A). Human males stand on two feet to urinate and hold their penis to direct the urinary stream.

Immediately a paradox becomes obvious. The human female is faced with an anatomical oddity. The urethra located anteriorly to the vagina has a posterior function; the vagina anatomically located posteriorly to the urethra has an anterior function (Table 2). In the human male, anatomy and function correspond (Table 3) but do not in the human female.

Table 2. Anatomy and Function in Human Female

Organ	Location of Organ	Location of Function
Anus	Posterior	Posterior
Vagina	Middle	Anterior
Urethra	Anterior	Posterior

Table 3. Anatomy and Function in Human Male

ORGAN	LOCATION OF ORGAN	LOCATION OF FUNCTION
Anus	Posterior	Posterior
Penis	Anterior	Anterior
Urethra	Anterior	Anterior

Consequences of Posterior Urination in the Human Female

To summarize this complex subject, let us say that a woman was able to copy man in terms of anteriorization of coitus but not in anteriorization of urination. It would be wrong to think this is trivial and of no consequence; the truth is that there are far-reaching consequences on human behavior: animals are differentiated as males and females only in the mating season, but humans are differentiated as males and females whenever they urinate.

My feeling is that since a girl takes a different posture to urinate than a boy, the mother teaches her daughter to urinate privately thereby conveying the message that there is something different or even "wrong." Thus, functional privacy is acquired. A simple and nonhidden procedure of urination in the male versus a complex and hidden procedure in the female is a constant reminder to women that there is a functional and anatomical difference. Women use a toilet seat both to urinate and defecate. Men use a urinal to urinate (it is easier) and a toilet seat to defecate; at least this is how it is in men's restrooms that are not unisex. In a household with the same toilet facility, there is no urinal and men use the same facility as women.

Urination as a posterior function in the human female and an anterior function in the male remains a hard problem to solve. Urinating privately only hides, not solves the problem. Let us mention that posteriorization of the urinary function is not performed by all women. It is not unusual to see teenage girls slightly spread and bend their knees after pulling down their panties and try to urinate forward like boys. Some modern female toilet rooms now include urinary bowls similar to those in men's restrooms. All a woman has to do is pull down her panties, apply the front part of her genital area against the bowl, and urinate like a man. Why these variations do not really take hold will be discussed later.

As noted previously, every part of the female body has been considered beautiful and sexually attractive. Yet, when it comes to the genital area (mons, labia majora and minora, clitoris, vaginal and anal entrance), there is absolute silence. I never heard of a woman praised for the beauty of her intimate sexual organs. At first thought, such an observation seems to be highly indecent. On further thought, however, it is in the genital area that the sexual act takes place and one may, therefore, wonder why consideration is not given for sexual attraction to the genital area (as animals do). As a gynecologist, I have dealt with women who complained about the esthetic appearance of every part of their body and discussed how to improve it by plastic surgery, or otherwise. I never heard a woman complain about the esthetic appearance of her external genital area. The truth is that all genital areas more or less look alike, and it is not in this area that a woman would like to be different from other women. While they do not mind being different from each other from the top of their head to the tip of their toes, women are more interested in making the genital area an inaccessible mystery, rather than exposing it. When I say women make their genital area an inaccessible mystery, I mean inaccessible to men, because among themselves women may freely discuss their genital appearance and even make comments on male genitals. Animals are more "direct" in that matter, and the sexual interest of the male is exclusively directed to the genital area of the female to which he is attracted by sniffing.

Let me immediately state that the above remarks are not general. The new "OUR BODIES OURSELVES" generation freely inspect their genital organs, although a mirror is necessary.

Men also become involved in hiding their genital organs, although to a lesser degree. The human male can be involved with his genital organs, although most of the time in a primitive, ordinary, and even vulgar way.

Men and women have always wondered why their penis and vaginal entrance are located so close to the anus and urethra, resulting in the sexual function being so close to defecation and urination. Certainly there could not have been a more awkward location. Women, because they may be more sensitive about it, have tried to remedy this difficult situation by distracting sexual attraction and stimulation away from its real anatomical location. In the particular sequence leading to the sexual act, it is only in the very last stage of the sexual process that the human female partner finally allows the sexual sequence to end up in its proper location (vaginal entrance and cavity). In no other mammal except in humans is a male attracted to any other part of the female body but the genital area. The human male would not mind direct access to the female genital area, but the human female would not allow it under any circumstance; she always requires that, first, other parts of her body be sexually attractive, then, the genital area. My impression is that if the vaginal orifice was not located between urethra and anus in humans, there would be less need for privacy in exposing the genital organs and more direct sexuality. Of course, there are other reasons for privacy that will be discussed later.

There is another major difference between animals and humans concerning sphincter control. In quadrupedal posture of animals, sphincter control of the bladder (urethral orifice) and rectum (anal orifice) is no problem because the weight of rectal (feces) and bladder (urine) contents are not directly against the respective orifices (anus or urethra) since the rectum and bladder rest over the anterior abdominal wall. In quadrupedal posture, the function of the anal or urethral orifices is just to open and close but *not to hold* stools or urine. But in erect posture, the weight of feces and/or urine is directed against the respective orifices, so that the function of the sphincter is not only to open and close as it previously was, but also to *hold* the contents that are now pushed against the respective orifices. When nonhuman primates use erect posture, they are well known for their urinary and fecal incontinence. This was a serious elimination problem that the early hominids (australopithecines) were faced with when they adopted erect posture. Therefore, the hominids had to build up stronger sphincters (Abitbol, 1988a,b) to prevent spontaneous fecal and urinary evacuations. Humans have developed powerful sphincters to hold the visceral contents while in erect posture; animals because of their quadrupedal posture do not.

Feces and their Repulsive Odor

The offensive smell was a problem that started with the early hominids and continues to be a problem for present-day humans. This problem is more prevalent among women than men, it is taught from a young age that fecal smell is offensive and expelling gas in public is to be avoided. This concept of offensive smell presents tremendous cultural variations. It always has been my impression that one of the reasons people in some cultures wear multiple layers of clothing all the way to their feet is to "attenuate" the odor (and noise too) when expelling flatus from the anus and "attenuate" other body odors. The repulsion for smelling or touching feces is extended to a similar repulsion for urine odor and contact. The latter is very pronounced in some Asiatic cultures.

This earthy discussion becomes difficult and embarrassing when the reason for the repulsive smell of feces and urine in humans is questioned. How acute the olfactory function was in the early hominid (Lucy) is unknown, but considering the similar brain anatomy of the present chimpanzee, we can assume that at the beginning or before the hominid lineage, the sense of smell was just as acute as in the present chimpanzee. From the evolutionary point of view, smell was our first sense and, according to Leakey (1965) and Leakey and Leakey (1964), our ancestors must have had a strong body odor, much stronger than present *Homo sapiens*. Certainly, fecal and urinary odors were included in general body odor. Animals do not consider feces (or urine) an obnoxious odor as do humans. Feces and urine are used in delineating animal territory. In fact the smell of feces and urine, just as much as pheromones, was helpful for the male in locating the genital area of the female in heat. I have no quarrel that it is the scent of pheromones that stimulates and attracts the male to the female genital area, although in some species the glands producing pheromones are not located in the genital area. However, I maintain that the stimulated male is attracted to the female genital area by the combined scent of pheromones, feces, and urine, because for millions of years their scents have been combined and they reinforce each other in producing sexual stimulation and attraction. What is certain is that the repulsion for the smell, and even sight, of feces is purely human and has, therefore, appeared somewhere during the hominid evolution.

In the beginning, the following scenario prevailed. Early in hominid evolution when erect posture appeared, the pheromones followed erect posture and moved anteriorly but fecal smell stayed posterior. One can imagine Henry's confusion, bending down and having to turn around Henrietta to sniff pheromones anteriorly and feces posteriorly. Henry's anger must have been great as he was used to sniffing everything posteriorly. Evolutionary loss of pheromones solved the problem, at least partly. There were no more pheromones anteriorly, but the smell of feces persisted posteriorly. One can speculate that the disappearance of pheromones made the odor of feces and urine more obvious and therefore contributed to Henrietta being sniffed more posteriorly than anteriorly since there was nothing to sniff anteriorly anymore.

We can hypothesize that to distract male attention from any posterior attraction and to eliminate sexual scent stimulations altogether, the feces, which are posteriorly located, acquired or rather were tagged as a bad smell. No male would look for sex in that area anymore. Since defecation is biologically necessary, the smell of feces could not be eliminated the way pheromone production was, so the smell of feces became taboo and bad.

Evolution of Defecation and Urination

It fell upon Henrietta and Henry, the two extinct hominids who underwent a three-million-year voyage of evolution, to go through this anatomical and physiological metamorphosis. When Henrietta (and Henry) stood up, the scenario started to change and became complete somewhere during the hominid lineage. Defecation is still a posterior function and the sexual function is now anterior for both sexes. Only urination posture differentiates both sexes, anterior in man and posterior in woman.

Henrietta and Henry at the beginning of the hominid evolution must have had difficulty in terms of urinary and fecal control since they lacked powerful sphincters. Careful investigation of Lucy's pelvis shows that the epiphysis for insertion of the anal sphincter is different from humans (Abitbol, 1988a, b), and this would imply incontinence. Imagine Henrietta standing up dripping with feces and urine, and pheromones in the process of disappearing. Only feces and urine would be smelled. How was this problem which was in dire need of correction dealt with?

So far the sequence of events for Henrietta in erect posture are anteriorization of coitus, posteriorization of urinary function, and loss of pheromones. It appears that the final result of posteriorization of the urinary function was not to the satisfaction of modern female *Homo sapiens*, and I hypothesize that she would have preferred anteriorization of the urinary function similar to the male *Homo sapiens*. The only solution was to make all these functions private, hidden from exposure.

The center of sexual activity in the animal female, who is approached posteriorly, is in the perigenital area where pheromones are assumed to be the mode of attraction. The human female is not at ease with this because it is posterior and close to the urogenital area with its specific olfactory characteristics. Therefore, she wants another location to be the center of sexual attraction that is anterior (similar to males) and separate from the unattractive uroanal area. We will see later that the new center of attraction is represented by the breasts and the face. To the naive question, "Why women do not expose their genital area?" my naive answer is simple. "It is because they also would be exposing the orifices of urination and defecation." Of course, there are other reasons, as will be discussed later. Natural selection favored females without pheromone production, but it was impossible to get rid of feces production. Therefore, all Henrietta could do was hide any fecal odor as much as possible, move the center of attraction higher up, i.e. face and breasts, and this is what she has been doing ever since.

No field researchers have reported any primate being repulsed by the smell, or even taste, of feces. In fact, Fossey (1983) described in great length the enjoyment of the mountain gorilla eating its own feces. Coprophagy is widespread in the animal world. The problem with coprophagy is not so much the health hazard (which is overly emphasized), but the irrational repulsion to it. It is also possible that our decreased olfactory power blurred the "pleasant" and "informative" part of fecal odors, leaving only the "unpleasant" and repulsive" notes. It is important to repeat and emphasize that the repulsive and obnoxious smell connected with feces is specific to *Homo sapiens* and therefore appeared during the hominid evolution. This may seem very surprising since the olfactory sense has markedly diminished in humans and, therefore, one would expect the olfactory perception of feces to be less intense in humans than in animals. Yet, the opposite occurred. The explanation becomes clear if one remembers that when the animal male sniffs pheromones in the perigenital area, he sniffs feces at the same time. Women do not want any olfactory sexual attraction toward the anogenital area.

The fundamental consequence is that during evolution of human sexuality, females have ruthlessly separated sexuality from urination-defecation, and considered them completely different from the anatomical as well as physiological point of view, in spite of their very close anatomical location. The result is an important aspect in women's lives. They dress and behave in general as if there were no rectum-anus or bladder, and they perform these functions in the strictest privacy.

Why Darwin's discovery that we descended from animals was so shocking now becomes obvious. By reminding humans of their animal origin, he was reminding them of similarities to animals in terms of urination, defecation, copulation.

Other Human Sexual Practices

The female anal orifice for sexual purposes is occasionally used as a substitute for the vagina. After decades of practice in obstetrics and gynecology, my observations in this matter are very simple. Some of my patients mentioned this practice, but once it became known that anal intercourse was the most common mode of AIDS transmission, it practically disappeared. However, when women occasionally practice this mode of sexuality, it is mostly at the request of a male. This practice is frequently painful to the female who does not enjoy it, and no woman ever told me she reached a sexual orgasm that way. I vaguely recall a couple of patients who mentioned some interest in anal intercourse. They were women who were having difficulty in reaching sexual climax and it was suggested in some women's magazines or by some "specialist" in sexuality to try anal sex. The interest was always temporary. The fact that anal sex is essentially a male request deserves special investigation. In her treatise "Sex in History," Reay Tannahill (1992) mentions anal intercourse as a mode of birth control in ancient (Greek) and primitive cultures. In fact, in ancient Greece, anal sex was a common mode of birth control.

Oral sex has provided some partners with a new field of sexual enjoyment. Concerning this subject, I will just mention that the "distance" from the anal orifice may be one of the factors responsible for this practice. This would be more evident with fellatio (mouth-penis), but to a lesser degree with cunnilingus (mouth-vulva). These practices are also reminiscent of the animal sniffing/licking.

The above discussion, in no way, is meant to be judgmental, although it may be embarrassing to some. Presently, some kind of reorganization and even "revolution" against these taboos and rituals is taking place. I believe that it is by going to the root of the problem, that many of these beliefs and misbeliefs can be properly evaluated.

Before the Dreadful Discovery

The three previous chapters have dealt with physical, anatomical, or biological transformations. Emerging intelligence made the anatomical differences between the two sexes even more obvious and they were reinforced by the advent of erect posture. At this point, some important definitions are necessary, such as exactly what is a male and a female. Biologists define each gender by three degrees of sexual characteristics; in this study a fourth group of characteristics will be added.

The primary sexual characteristics are essentially genetic in nature. The two sexes have different chromosomal cell patterns easily recognized with microscopic examination. The sex chromosome, or X chromosome, is present in the male and absent in the female (Y chromosome).

The secondary sexual characteristics involve copulation and reproductive organs: ovaries, tubes, uterus, clitoris, and vagina in females; testicles, prostate, seminal vesicles, and penis in males. They are either hidden within the body in the female or externally located in the male.

Tertiary sexual characteristics are anatomically unrelated to copulation or reproductive organs and located externally in male and female: for female, hair distribution, breasts, curved body contour, shorter stature, weaker musculature, high-pitched voice, etc.; for male, different hair patterns, flat breasts, angular body contour, taller stature, stronger musculature, low-pitched voice, etc. These tertiary sexual characteristics are not absolute but are basically present in all men and women with minimal individual variation.

To these, men and women added quaternary characteristics to further differentiate the physical appearance of the sexes and also to reinforce their natural sexual appearance (Chapter VI). Although there are vast cultural variations, it is believed that (1) women do more to enhance feminine physical appearance (clothing, makeup, hair style, etc.) than do men; (2) males have more authority, think more than feel, repress emotions, while females are more passive, feel more than think, freely express emotions, and more involved in parent-infant bonding; and (3) male activity is mostly outside the home and female activity is mostly within the family circle. All these quaternary characteristics are arguable and subject to continuous changes.

In humans these added sexual differences are so deeply ingrained that they become a way of life. Parallel to the evolution of anatomical sexual characteristics, there is a progressive emergence of feminine and masculine traits or modes of behavior. What makes *Homo sapiens* male or female is certainly anatomical, but more so behavioral. There are so many variations that, for practical purposes, masculinity and femininity will take an aspect specific to each individual. This is the fundamental difference between animals and humans in terms of sexuality.

How the Dreadful Discovery Occurred

When we stand up and face another person, we see the front of another person's body from head to toe; we see our own image. It is while inspecting the front of the body in erect posture that the *dreadful discovery* (absence of genitals in the female and fear of their loss in the male) is made. Henrietta can hide the absence of genitals with pubic hair, but eventually she has to realize that, compared to Henry, something is lacking behind that pubic hair.

It is possible that the *dreadful discovery* started with the difference in urinating postures; it made the absence of penis and testicles more obvious. In addition, the sexual act when performed anteriorly, made it obvious that the male had a protruding organ and the female had a cavity. This became even more obvious when Henry had an erection and his protruding organ penetrated the female cavity.

This discovery could have happened any time during the hominid evolution, either at Australopithecus or Neandertal level, or any level in between. How much intelligence was necessary for this discovery, or if this discovery was made suddenly or progressively over millions of years, is not known. The only requirement for this discovery to be possible was full erect posture and front-to- front sex where the genitals were fully exposed. Only in erect posture could the hominid inspect his/her own genitals or the genitals of the sexual partner by looking at them and touching them. The other requirement was that a certain amount of intelligence was needed for the *full* realization of the *dreadful discovery*, especially when one considers all the consequences that followed.

Let us suppose that Henrietta and Henry, both with perfect erect posture, are inspecting their own and each other's genitals. This would be a normal occurrence in view of the natural curiosity expressed by all primates. To Henrietta's horror, she discovers she has no external genital organs, while Henry displays two testicles and a penis that can be enlarged as a result of erection following simple touch and prior to coitus. Through thorough inspection and comparison of her body with Henry's, Henrietta discovers she has only a cavity represented by the open space between the two labia, the vaginal cavity. Upon further inspection of the bodies of all other Henriettas around, she makes the same observation. On repeated inspections, it is obvious that the labia and the clitoris are poor substitutes for testicles and penis. The final conclusion is that something is missing in Henrietta and there is something additional in Henry.

This *dreadful discovery* represents the initial event that separated the human male from the human female, that created man and woman, Adam and Eve. Individuals of Henrietta's and Henry's generation recognized that they could be divided into two categories, the "have" and the "have not." The human species is still divided into two categories that may proceed to eternal conflict and attempt a union in spite of or based on these differences. This perception of sexual dissimilarity is completely different from that of animals or animal-like species that preceded Henrietta and Henry. If animals perceive sexual differences, it is only when they are in heat and during the sexual act itself. With Henrietta and Henry it was a permanent anatomical difference, not exclusively related to the time of the sexual act, although it made the difference even more obvious.

In the beginning when the australopithecines first stood up, it was very likely that Henrietta did not have enough intelligence to be aware of the *dreadful discovery*. There is no *dreadful discovery* with apes, although every present-day ape has a natural curiosity and a tendency to manipulate and explore all foreign objects. It is highly likely that sometime at the beginning of the *dreadful discovery*, Henry became aware of what he had and proudly exhibited the difference, thus reinforcing Henrietta's frustration.

The many natural consequences of the *dreadful discovery* have marked mankind forever. What happened afterward is almost folklore, with dramatic and heroic moments. The separation of the sexes had begun and would determine not only hominid sexual life but all life thereafter. A new life began with sexuality no longer under biological but conscious cerebral control.

We now have the four major elements on which human sexuality was built: ventral-ventral sexual posture, loss of pheromones, visible male genital organs and not visible in the female, and control of sexual behavior by the cerebral cortex instead of sexual hormones. What role these elements played, how the sexual revolution occurred, and how Henry and Henrietta relate to each other had to be completely revolutionized. Achieving full awareness of these differences and overcoming them represent the major elements on which human sexuality, or rather the whole human culture, has been built.

The different and successive reactions of Henrietta and Henry to the *dreadful discovery* originate as much from Henrietta's feeling that she is "inferior" because she thinks she is missing something as from Henry's superior feeling because he thinks he has something additional (Freud S, 1930; Freud A, 1936).

Women, being of the same intelligence as men, are not going to react indifferently to male "superiority". Henrietta will react in many ways: I am like Henry; better than Henry; different from Henry in a better way. When dealing with sexual differences, one often ends up in establishing differences with everyone else, including individuals of the same sex. Women will approach these problems not only as a group but as individuals, each in her own unique way. For instance, a girl picking a flower, putting it in her hair, and looking in a mirror or at her reflection in water to see how pretty she looks, has been repeated thousands of times over the last three million years, but each girl will choose a different flower and place it differently. Female sexuality is both general and individual. This flower girl is no different from today's woman who purchases an exclusive and expensive dress just to be unique. Men follow a similar pattern but to a lesser degree, as will be seen later.

First Reaction, Hiding

The first reaction of Henrietta and all subsequent women to the *dreadful discovery* is to hide her genital area or even herself. At the time of the initial shock, Henrietta behaved the way all creatures do when they witness or are part of a shocking event. Imagine a modern teenager who is preparing to dress in the privacy of her room. She is completely naked and a stranger suddenly opens the door and enters her room. Her reaction is all too familiar. The young girl screams, flexes her thighs and knees slightly to retract her perigenital area, and tries to cover it with one or two hands. She will also try to hide her breasts and, depending on the angle of exposure, she may put one hand behind her to cover the perianal area. The whole episode ends up with her grabbing a sheet to cover her perianal and perigenital areas and her breasts. If it had been a naked male teenager, his reaction might have been the same, but milder. This modern reaction is not necessarily that of Henry and Henrietta during the initial *dreadful discovery*, but it certainly can be a basis for comparison.

Therefore, the first reaction, an ostrich reaction, is to hide the perigenital area, and the pubic hair was helpful in this matter. Evolution may have been favorable to those women with abundant pubic hair, but not to those with hair on other parts of the body. The rule in the female gender is absolute for all hominids following the *dreadful discovery*: the perigenital area is never to be exposed, and this rule has never changed. Anything that is handy will do to hide it, whether it is a fig leaf, animal skin, or any kind of clothing. In fact, hiding it is so common that females even hide it from each other, but more so from males. With *Homo sapiens*, the rule is still prevalent but there are vast cultural variations. One extreme is a small "cache-sex" placed anteriorly and sometimes posteriorly, another extreme is hiding the whole body, except for the eyes and hands with clothing, as in some Moslem cultures. There are even some cultures today that hide women behind walls and never expose them to public view.

In my gynecological practice, I have seen many teenagers brought to my office by their mothers for gynecological evaluation at puberty. Their embarrassment at a gynecological examination sometimes can be extreme. I vividly recall cases where the teenager declared that, no one, not even her mother, had ever seen her genital area. I even recall a girl who stated she had never seen the genital area of another girl. Pediatricians who treat adolescents confirm this opinion about girls, but not about boys. Of course, this is not a general rule and there are many variations. The hiding game starts within the family circle, between siblings and between parents and children. However, this first reaction to hide her genitals is temporary; it just gives women time to think, plan, and react.

Second Reaction, Making it a Mystery

The successive hominids with increasing brain capacity and more complex use of visual, auditory, tactile, and manual functions developed a more elaborate and subtle strategy to cope with the *dreadful discovery*. These strategies show the development and formation of the male and female personality. With emerging intelligence, subsequent male and female hominids were capable of much more than hiding.

This hiding is very often so extreme that one may rightfully wonder why it is so intense. What made me come to this conclusion is that I have personally interviewed male adolescents and, to my surprise, I learned that many of them have never seen a complete view of the genital area of a girl, even though some of them have touched that area in the process of sex play. I found it to be an almost absolute rule among siblings. Most girls manage to peek at the genital area of their brothers, especially younger ones. Girls six years of age change the diapers of their young brothers/sisters or any other infant. One of the reasons is that they want to verify the *dreadful discovery*. Yet, I barely recall a case of a teenage boy having seen the genital area of an older sister, although many have attempted to do so. The answer to this strange behavior was given to me by some male teenagers whom I asked to describe how they imagined this hidden and untouchable area to be. Their fantasy is unbelievable; they imagined a hidden and mysterious organ, an inner sanctum, having a magic effect on whoever touches and sees it. Some teenage boys obsessively fantasize and dream about it, and looking at pictures in sex magazines barely satisfies their curiosity. The bizarre aspect of these fantasies is that they are present in boys who have indulged in sexual activity as well as in those who have not. Their fantasies are mostly directed to girls who make themselves inaccessible to any kind of relationship. Pushed to the extreme, some boys fantasize that the female sexual organ is like the mouth of a monster with sharp teeth, "vagina dentata," (Paglia, 1990) who will bite and eat anything that comes too near (penis, hand, etc.). The vaginal cavity has become a mysterious hidden cavity, the second reaction of Henrietta to the *dreadful discovery*. The power to create and procreate life in the female genital organs and the expulsion of the fetus through the vagina further intensifies the mystery.

Very early in life girls are ordered to cover themselves, and boys are taught not to look, and vice-versa. This technique has been maximally perfected since Henrietta (and Henry) have had millions of years to improve it. Access to this mysterious cavity is like access to any mysterious location, requiring a whole preparatory ordeal to reach it. Since the anatomic appearance of the external genital area of any female is ordinary and without a specifically attractive feature, it is important to cover up this area, thereby creating something mysterious and barely accessible.

Each step in the final access to the vaginal cavity is complicated and mysterious. When Henrietta stood up, the female body lost its neutral composition and became one of positive and negative zones. Among the numerous aspects of femininity, one of them consists of how to glorify what is shown and to mystify what is hidden, and to alternate between what is shown and hidden. A sixteen-year-old girl, when asked her opinion about nudity on the beach responded spontaneously by saying, "No way, it leaves nothing to the guys' imagination."

Although the human female has lost estrus and all the advertising of sexual receptivity that goes with it, she makes every effort to appear externally sexually attractive. Her real sexual receptivity and desires are now concealed and controlled. At least the animal male has the advantage of identifying the sexually receptive females and is sexually attracted only to those. The human male has the heavy burden of determining whether or not the human female is desirous of sex and will respond to the proper stimulation or courtship. The female sexual partner is now inscrutable and nondecipherable. A mystery!

Women, more than men, spend a great amount of time in private when dressing or undressing, washing, etc. Men are less private, and these community functions are accepted more by men. Men and women expend a tremendous amount of energy on this hiding game. Women's cerebral energy is spent on hide'n'seek and seductive behavior while men spend a good part of their lives "running after skirts" and what is mysteriously hidden under those skirts.

To reinforce the mystery of the genital area, the female would prefer that sex be performed in privacy and in darkness. Privacy of sexuality means access limited to only one person and not for public display. Other reasons for sexual privacy will be discussed later.

Hiding and mystery were only initial reactions to the *dreadful discovery*. Henrietta needed a more positive and substantial mode of sexual attraction. Rather than make a mystery of the absent organs, she needed her own protruding organs since the female genital organs were at a disadvantage because of their location. Other organs or modes of sexual attraction that could be elevated and anteriorly located certainly would be more effective. This would allow Henry to remain standing instead of bending down to sniff pheromones. Besides, hiding and mystifying the perigenital area makes it no longer a zone of primary sexual attraction.

The search for a mode of attraction to replace the "absent" organs in the perigenital area consisted of three stages. The initial stage is superiorization at the breast level, the next stage is superiorization to the female face, and finally, generalization to the whole female body.

In using different parts of her body, Henrietta's use of the hide'n'seek technique has worked well so far. She will expose some organs and hide others in an attempt to create a mystery. This will be determined by personal and cultural variations.

Third Reaction, Initial Superiorization or Appearance of Breasts Mammary glands are common to all mammalian species. Outside of the lactation period the mammary glands are quiescent and almost atrophic, and during pregnancy they hypertrophy under the influence of the hormones of pregnancy. The mammary glands in the nonhuman mammals do not present any enlargement visible from the outside, and one would not recognize the sex of an animal by the external appearance of the chest. Even during lactation the animal mammary glands are not very enlarged. The only exception is the cow, which has been producing milk for human consumption for centuries.

One of the most fascinating aspects of the evolution of human sexuality relates to the appearance of breasts in the human female. Unlike other mammalian species, the human adult female presents permanently enlarged organs called breasts. Breasts are characteristic of human females and they are an important anatomical element differentiating females from males, besides the external genitalia. The human breasts contain mammary glands identical to the animal mammary glands and also anatomical elements that give them their permanently elevated and round shape. Under microscopic examination, breasts are essentially composed of fat and connective tissue abundantly present between the minute mammary glands. The nonhuman species have organs composed *only* of mammary glands. It is the presence of fat and connective tissue that makes the human mammary glands become breasts. Breast size varies greatly among females of the human species; large breasts do not necessarily signify abundant mammary glands producing more milk during the lactation period. Conversely, small breasts do not necessarily produce less milk during lactation, they just contain less fat and connective tissue.

The present culture's approach to the human breast is indeed puzzling, and can be divided into two groups.

The first group is composed of scientific people (mostly biologists, physicians), scientists of various disciplines, and even religious authorities. They regard breasts only as mammary glands performing their lactation function. Without exception, they explain breast enlargement as necessary for milk production and nourishment, which indeed would require large breasts. It is rare that a scientific article mentions breasts in any other respect but lactation as a primary function. The only exception may be plastic surgeons who enlarge or reduce breast size for purposes other than the lactation function.

The second group is composed of nonscientific people who, although not denying the lactation function, consider breasts a sexual organ. Numerous medical or biological books and scientific publications on normal and abnormal breasts do not mention any sexual connotation. Reciprocally, numerous women's magazines and books on human sexuality mention breasts only as sexual organs; they barely discuss the lactation function except to mention that pregnancy and lactation distort their shape and make plastic surgery imperative. I mention a single religious exception: the Song of Songs of King Solomon describes breasts as sexually attractive and not in connection with their nourishing function.

Psychiatrists and psychologists have a variant attitude. They state that during breast feeding, the infant developed some kind of atavistic and physical attachment to the maternal breast subsequent to the nourishing function, and this persists subconsciously during adult life.

These classic views are open to criticism.

If fascination for the human female breast was an atavistic and unconscious remnant of physical contact and nourishment during breast feeding, why is it that nonhuman mammals, who are highly capable of developing conditioned reflex behavior and some kind of attachment to that part of the body, do not remain attached to the maternal breast? Yet, no one has ever reported this behavior in any mammal, not even in higher nonhuman primates. It is not denied that breast feeding and physical contact with maternal breasts leave a permanent mark on the human psyche, but it is only a partial explanation. It is my personal impression that breast feeding has markedly diminished in our culture and yet breast attraction is not decreasing. Fascination with female breasts is the same for those humans who were breast fed as those who were bottle fed.

Only human males are fascinated and constantly stimulated by female breasts, though females are to a lesser degree. Yet, both male and female infants are, or are not, breast fed equally.

The function of the human breast as a nourishing organ is fully respected and praised and whether women breast or bottle feed is not the subject of this study. Essentially, the points argued in this study are whether or not human female breasts have enlarged to further fulfill their nourishing function, if there is a connection between enlarged breasts and sexuality, and if sexual attraction to the breasts should be considered a pathological deviation.

It cannot be denied that the female breasts in humans function as nourishing organs and also as sexual organs. Nursing women frequently have told me that strong and frequent suckling of their nipples acts as a sexual stimulant. This feeling is so embarrassing for some that it contributed to their giving up breast feeding. Women may not be as shy in exposing their breasts as they are in exposing their genital area, but they certainly are shy with both, and will cover their breasts as quickly as their genital area. Next to facial appearance, the breast is an essential element of female beauty. No woman considers herself sexually complete until she has adequate size breasts. The breast is an integral part of sexual stimulation, and all sexual partners know that the breast stage of sexual stimulation is mandatory. This sexual function of the breasts, not present in animals (not even in higher nonhuman primates such as chimpanzees, gorilla, orangutans) and full blown in *Homo sapiens*, obviously appeared during the hominid evolution. How the female breasts appeared and became enlarged, and how they became a sex symbol during the hominid evolution will now be considered.

Having made and survived the *dreadful discovery*, Henrietta desperately studied her whole body to find something protruding, as anatomically distinctive and as big or bigger, than a penis and testicles that could be proudly exhibited. I suppose that, consciously or not, Henry's exhibition of his organs and an overt feeling of being complete, contributed to Henrietta's increasing frustrations. As an expression of frustration, and with her emerging intelligence, she wants something different and more attractive than a penis and/or testicles. The penis being a straight organ and producing pain when penetrating, the female wanted something gentler and softer to the touch, preferably an organ anteriorly and superiorly located.

It is not known how many millions of years the search for this sex symbol has gone on or how many different choices were available. What is known is that the final choice was the two breasts. It could not have been a better choice. Indeed, the two breasts have a long history as a source of attraction and fascination for Henry (and Henrietta) since birth. Indeed, this organ provided nourishment, warmth, protection, and intimate contact with the mother. This intimate physical contact became more meaningful during the hominid evolution because of the developing sense of touch, a sense almost nonexistent in animals.

Breasts have no sexual meaning in animals, even in the higher primates. Extensive observations of sexual activity in various animal species do not show any kind of manipulation of the mammary gland area that is even remotely sexually significant. The animal infant approaches the maternal mammary glands and nipples only for suckling, and once nursing is finished the animal infant no longer approaches the mammary glands. Adult animal males have no interest in that part of the female body whether or not in a state of sexual attraction. The sexual aspect of the breasts is a purely human acquisition.

Indeed, the breasts have some characteristics that make them sexually attractive. Some woman may fantasize that the milk-producing breast is similar to the sperm-producing penis. Gentle breast manipulation, especially around the nipple, produces a hardening of the nipple that could remotely be a reminder of penis erection. Breast manipulation as a preparation to the sexual act is appreciated by both human sexes for several reasons: sexual stimulation for both, and reminder of breast feeding or a lost paradise. Therefore, besides the fact that they are anteriorly and superiorly located, Henrietta had other reasons for selecting the breasts as a positive mode of sexual attraction, such as rubbing them against the male body during the sexual act. Henrietta, already an expert in the hide 'n seek process of her perigenital area, has no problem with inclusion of her breasts in this act. She will hide her breasts with bras (breasts often need support) that add mystery and enhance them so that they will more or less protrude; this is done under the guise of shyness and prudishness. Breasts can be manipulated around-the-clock, which is an advantage over penis erection. Also, the fear of breast castration compared with the fear men have of losing their penis and testicles is very limited. Natural selection favored Henriettas with breasts permanently enlarged by fat deposits and not small breasts.

To understand the phylogeny of permanent breast enlargement, let us look at its ontogeny in humans. One can understand the eager expectation of the human female teenager as she watches the slow growth of her burgeoning nipples and breasts as a positive, protruding sexual organ that can be hidden behind bras or exhibited and that can stir imagination and fantasies in boys. The hope is that they will grow to "adequate" size; the fear is that they will remain small in size.

According to Freud, awareness of sexual differences is seen at a much younger age (four to five), characterized by the *dreadful discovery* in young girls and the "wonderful" discovery in boys. Between four and eleven years of age, girls live in a state of constant frustration because of the *dreadful discovery.* It is only at the age of eleven that the third reaction begins with breast formation and menstruation, and positive elements of female sexuality appear, which can be used as substitutes for the *dreadful discovery* and the absent genitals. Breast formation is a welcomed relief after prolonged anticipation. Women never really hide/ expose their breasts the way they hide their genital area, but do so more than males who hide/expose their genital organs. A flat-chested woman may consider herself less feminine. Women can easily play the game of hide'n'seek with their breasts: how much or how little to hide behind bras or exhibit and how to modify their shape through type of bra is an art quickly mastered. Breasts are among the sexual features of which females are the most proud, second only to facial appearance.

So far boys have been playing the "hide'n'seek" game with their genitals more or less hidden underneath their clothing. But now girls can do the same with their breasts underneath their bras! This is essentially a search or an exploration by the two youths of the opposite sex who are sizing up each other in preparation for full cooperation.

Another way to understand the phylogeny of breast enlargement in human females is through primitive art. Some African sculptures on this subject are revealing and even very suggestive. One sculpture of two elongated breasts with the nipples protruding gives the impression of two huge penises, and the milk coming from them easily can be compared to sperm. Two huge breasts also can be compared to two huge testicles. The general feeling this sculpture imparts is one of aggression. The female in this sculpture is slightly flexed at the thighs thus retracting the genital area and hiding the absent male genitals. The perception of hostility and fear is even better demonstrated by an antique Roman Goddess with multiple breasts that easily match the aggressive appearance of any testicles or penis in erection.

I do not mention the nourishing function of breasts because after forty years of practice as an obstetrician and gynecologist, I have almost never heard it mentioned by any patient. I have dealt with patients who talked about breasts that were (1) too large and how to reduce their size, (2) too small and how to enlarge them, (3) unequal in size and how to equalize them, (4) inverted nipples and how to evert them, (5) hair around the nipples and how to remove it, etc. The only time milk formation came into discussion was when the patient requested medication to prevent or stop it. Most patients complain about the stretching and hanging of their breasts following pregnancy and breast feeding. Plastic surgery on the breast is much more popular than breast feeding. Breast feeding in humans is popular mostly among those who advocate it and much less among those who practice it. In a modern culture and despite efforts by advocates, only five percent of women who give birth, regardless of all the pressures and advertising, breast feed. This group practices breast feeding only for a short period and always in combination with artificial feeding. Providing that substitute milk is abundantly available, mothers in any part of the world prefer it to breast feeding. They consider their own milk supply erratic, and besides they become tied up with the newborn day and night. To some degree, fathers do not mind bottle feeding because they can also "give" the bottle. No scientifically conducted study has ever proved that breast feeding is superior to artificial feeding, providing the latter is done skillfully and under proper medical supervision. A baby, given a choice, would prefer a bottle because milk is more difficult to suckle from the maternal breast.

Concomitantly with the breasts and to a lesser degree, prominent buttocks have become a classic female sexual attraction. Some cultures consider flat buttocks a lack of femininity almost to the same degree as a flat chest. From the Venus figurines carved 30,000 years ago by ancient people from Europe, to the not so distant use of bustles and corsets, breasts and buttocks always have played a role in female attractiveness. Forsyth (1966) thinks that Victorian modesty prevented Darwin from writing explicitly about the virtually universal admiration of female breasts and buttocks. Ice Age carvers, inventors of bodices, bustles, corsets, and brassieres, and Hollywood producers and pornographers alike have always emphasized the sexual aspects and not the nourishing aspect of the breasts.

Fourth Reaction, Facialization

Facialization is the fundamental innovation of human sexuality. Obviously it started with face-to-face sex which allowed the sexual partners to look at each other's face. Today, the most important and maybe the most unique criterion for evaluating sexual attraction in a woman and a man is facial features. This is the concept of beauty. I always have difficulty explaining that visual beauty is unique to the human species and does not exist in animals. When I attend a dog or cat show, I have a hard time explaining to people that this is a show for humans not animals, and that when we award first prize to a beautiful-looking and well-groomed dog, we really give it to the owner for making the dog look good to human eyes. If dogs had ways to evaluate each other (and in their own way they do), the prizes would be based on some olfactory qualities.

The concept of beauty is specific to humans and arose during the hominid evolution; I call this process facialization. Why is it more appealing to have sex with a woman who has beautiful facial features than with a woman who does not? This may appear to be a stupid question. Yet, I challenge anyone to give me a reasonable answer. In this chapter I will describe how the center of sexual attraction has shifted from pheromones to facial features. Obviously, evolution, specifically sexual evolution, favored development of specific facial features or beauty in women and, to a lesser degree, in men.

The face of humans, even though it possesses the same anatomical components as in other primates, is basically different; it has the eyes and ears like other mammals, but their features are more elaborate. It also has a mouth (with tongue and lips) that has become an organ of speech. The face is less hairy in the human male and there is little or no hair in the human female. The jaws have retracted backward and the forehead moved forward. An important characteristic of humans is that in erect posture the head is easily held in equilibrium over the neck with limited or no contraction of the muscles of the back of the neck. This results in freedom from the concern of holding the head in equilibrium. Consequently, facial musculature can be used for facial expressions. More muscles appeared on the face and the whole facial musculature has a new fundamental function: the refinement of facial expression.

The use of facial features and expression fulfills an urgent need resulting from a vacuum of sexual stimuli. The face easily can fulfill this role for many reasons: it is fully exposed due to disappearance of hair; its mobility of expression increases because of more facial muscles and use of language; prominence of jaws is regressing in light of different eating habits and because teeth are being used less as a means of aggression or defense; nose is not as wide due to regression of olfactory function; and the forehead is expanding because of encephalization. With the appearance of language, which is the most important mode of communication, eye and ear contact becomes essential in humans. It is not so important in animals, even among higher primates. When a group of monkeys or chimpanzees is observed, it is easily noted that eye-to-eye contact is rather short. From now on Henrietta will do anything to draw attention (sexual or otherwise) to her facial features. Forget about absent genitals and pheromones, and let us concentrate on the face which can be looked at freely. Of course, Henry also finds advantages in facial expressions, and goes along.

At the beginning of the hominid evolution, Henrietta was still undergoing cyclical estrus stimulations or sexual heat. But, pheromones disappeared and males were no longer coming around. Henrietta felt confident in using facial attraction as a tool since she had a couple of million years at hand to test, retest, and perfect this new and surprisingly effective mode of attraction. The process of facialization must have been long and arduous. How the present features formed themselves during the hominid evolution was certainly complex and elaborate, and explained by the important role played by the rise of visual, auditory, and tactile functions. Indeed, facial attraction is mostly visual but it also has auditory and tactile functions. When Henrietta spoke, she formed special facial expressions. With the regression of aggressive biting and evolving hairless body, touching Henrietta's face must have been more enjoyable. I suspect that the loss of hair first began with Henrietta's face, and also that the preponderance of the tactile function during the hominid evolution started with, and was parallel to, hair loss.

Henrietta first saw Henry's responses to her facial expressions during face-to-face coitus. With her rising consciousness, she was quick to perceive the potential of attractive facial features beyond and outside of coitus and has never stopped expanding their use. Hiding the genital area and exposing the facial area is the new Henrietta.

All the Henriettas and Henrys of the hominid evolution will learn to look at each other, with eye contact becoming predominant and used more freely; they will learn to listen to each other and to watch each other's facial expressions while talking/listening, and they will learn to touch each other's newly hairless skin. All the elements are now in place for the face to acquire special significance in hominid relationships. Near the end of the hominid evolution, facial features and expressions become the preponderant expressions of sexuality with unbelievable advantages.

Some of these advantages are: males can be kept at a distance through visual rejection or signs of non-receptivity; the sexual expression can be directed to a specific male and not to a whole group of males; and most of all, Henrietta can control her facial expressions to decide when, with whom, and how to be sexually attractive. Henrietta already has accumulated some experience in sexual expression when she redirected the sexual focus from the perigenital area to the breasts. She, therefore, will have no trouble in shifting the focus of sexuality to facial features. Facialization is complete and facial beauty is born. Beauty is not only the possession of special facial features but also the art of facial expression, which could be called charm.

The numerous facial muscles of pongids, although not as elaborate as humans, are rarely used. Pongids (chimpanzees, orangutans, gorillas) are capable of numerous facial expressions, mostly to express fear and anger (Life Before Man, Time-Life Book, 1972). Otherwise, the faces of all animals are practically expressionless (even during copulation). The reason they rarely use or require facial expressions is they have no use for it and communication is well established through pheromones and regulated through hierarchy. They have at their disposal enough rudimentary modes of relationship without it.

Humans communicate as a way of life and facial expressions have become an important mode of communicating feeling and emotions. One can see therefore the meaningful role played by sexual activity in the rise of facial features and expressions. I am convinced that it is during face-to-face coitus that facial features started to be perceived and facial expressions started to be exchanged between the two sexual partners. Fear, hate, pain, suffering, pleasure, satisfaction, ecstasy, orgasm, and so on, can all be expressed through facial expression during coitus. Facial expressions became more useful and generalized to other aspects of human life besides sexuality. The biological woman (femina biologica) became a seducing woman (femina seductens) and her main tool was facial expression.

Facial expression and charm are both logical substitutes for depheromonization and a logical sequence of the *dreadful discovery.* Through beauty and charm, women have now established some kind of equilibrium to balance physical strength and whatever additional anatomy men think they may have. Some will go for a destructive confrontation, some for a hopeful union. Facialization will save women from being enslaved by men. They will attempt to dominate men through their facial beauty the way men attempt to dominate through physical strength. The two sides will learn each other's potential and capability, and also the dangers of confrontation and the advantages of cooperation.

Fifth Reaction, Generalization

By now, Henrietta has developed a tremendous amount of confidence and expertise in utilizing her full potential for sexual and all other purposes. She is now a real artist and, in fact, a discoverer and an inventor. She is a genius in all these respects. She is just as intelligent as Henry, but she is more motivated because of the effect of the *dreadful discovery*, which was more profound on her than on Henry. This makes her highly motivated, and she has more drive and determination to foster the hominid evolution. It is as much the evolution and birth of "*Femina sapiens*" (wise woman) that we are witnessing as the evolution and birth of "*Homo sapiens*" (wise man).

As one can easily imagine, she is not going to stop there. She was so successful using the breasts and the face as sexual attractions, why not use each and every other part of her body. Since she has bad memories of the perigenital and perianal areas she will avoid them.

The elements of female sexual attraction can be divided into two categories. First the hide'n'seek rule which basically is how to expose, flaunt, cover up, or hide completely each part of the female anatomy. Individual and cultural variations are many. The female who, following her customs, hides every part of her body except her eyes and fingers or the female on the beach who wears only a "cache-sex" over her pubic area, are really showing various forms of the hide'n'seek technique. Between these two extremes are women who will shift the length of their skirt to show more or less of their legs and thighs, and those who will expose more or less of their inner thighs simply by sitting/standing or crossing/uncrossing their legs.

How to modify the external appearance of specific parts of their body to make them more sexually attractive is the second category. Enhancing external appearances must have started late in the hominid evolution because it required more intelligence and different tools. The oldest procedure might have involved hair: combing it in different ways, keeping it long, cutting it short, or changing its shape. The next important step might have involved makeup or color, red on lips and black around the eyes to enhance them. Next, the addition of objects about the face: flowers in hair, rings in ears or nose, necklaces, tattoos, and other ornaments. Dental and plastic surgery are the most recent acquisitions. Each Henrietta wants to be unique in her physical appearance and sexual attraction, and still is searching for the ultimate tool (special perfume, lipstick, or nail polish) or sexual stimulus as powerful as the lost pheromones. We can see that replacing pheromones is still a continuous endeavor.

Most of all, each woman fantasizes she is highly individualized and unique, that she is the only one with a magical power. The question is how to discover or rediscover this power (pheromones) and manipulate it. Sexuality has moved from a biochemical, pheromonal, subcortical mechanism with its stereotype character to one that is now visual, auditory, tactile, neo-cortical, individualized, and whatever goes with it.

The whole female body becomes orgasmic, which means that almost every part of the female body can be used for sexual stimulation. Every part of the female body can, through sight and/or touch, sexually stimulate the male, and the male touching any part of the female body can be sexually stimulating for her. The result of stimulation worked to a lesser extent in the male body. The penis is really the only part of the male body that is sexually stimulated mostly by touch.

It is well known that neglect of physical appearance is a sign of serious emotional disturbance, more so in women than in men. Such neglect can be taken casually by men, but never by women. The reason is that the modification of external appearances is now part of nature in women. I know women who will never appear before anyone (not even their husband or children) except when properly dressed and with makeup. Some women even go to extremes, such as avoiding daylight because it exposes wrinkles, and living in a state of permanent semi-obscurity. As a gynecologist, one of the surest signs I pay attention to on deciding when to discharge my patients after surgery is seeing the patient on my morning rounds with her hair combed and wearing makeup; she has returned to normalcy and is ready for discharge.

Sixth Reaction, Additional Minor Reactions

In addition to the previously mentioned five major reactions to the *dreadful discovery* there are also numerous minor reactions. Mention is often made of the clitoris being an erectile organ in the female. The clitoris of the human female is the smallest among the primates. It is the reverse for the penis size of the human. During the hominid evolution, therefore, the penis progressively increased in size and the clitoris decreased in size. It is not difficult to reconstruct the events that surrounded these opposites in size. In erect posture, Henry fully exhibited his penis and testicles. It was becoming more and more obvious that sexual prowess is connected to penis size and performance, that Henry is proud of the size of his penis and attracts Henrietta accordingly. Natural selection, at least sexually, was in favor of a larger penis for Henry. Male nakedness is not unusual in some primitive cultures and Van der Post (1961) reported that in some of them the penis is in permanent erection. In some Polynesian tribes, naked males hold their penis with a string to enhance exhibition.

Henrietta, once she made the *dreadful discovery*, became more and more involved with the hiding process because, regardless of how big her clitoris was, it still was very small compared to Henry's penis, especially when erect. Besides, Henrietta was no longer interested in her genital area as a zone of sexual attraction (contrary to Henry) as she focused on other parts of her body (superiorization, facialization, etc.). The size of her clitoris was less significant and even an obstacle in the hiding process. Natural selection was, therefore, in favor of females with a smaller clitoris. In some primitive tribes, clitoridectomy (a female circumcision, which is really a clitoris resection) is still common practice.

If breast enlargement is the first sign to affirm female puberty, the second sign is menstruation. This periodic bleeding gave the mixed appearance of a bleeding trauma, something brutal and harmful, and completely opposite the gentleness and harmless innocence that is appearing more and more in women. Blood means fear, but also courage. Certainly, something frightening and mysterious is going on. This atmosphere of mystery was part of the Elusian mysteries described by the Greeks. Menstruation is never discussed aloud, it is always whispered between women and only in strict intimacy between mother and daughter. Under no circumstances would it be a subject discussed in front of a man, even a brother, father, husband. Not many women are aware of the physiological significance of menstruation, but they perceive mostly its psychological and mysterious impact. The first menstruation is a well-defined sign of puberty in females, while there is no such specific landmark in males. Males have to rely on such signs as pubic and/or axillary hair which is progressively appearing, while the first menstruation clearly defined the girl as a woman.

After breast development and menstruation, pregnancy is the third anatomic difference in the female body. By its size, a pregnant abdomen can match any male organ. It is my conviction, as an obstetrician, that the happiest moment in a woman's life is when she is pregnant, usually between four and eight months. Prior to four months, the fetus is not yet moving, there is danger of miscarriage, and the enlarging abdomen is not very noticeable. After eight months the size of the pregnancy becomes cumbersome and interferes with many functions, fear of labor and delivery becomes an obsession, and there is a growing concern for the coming child (congenital malformations, traumatic birth, socioeconomical considerations, etc.). But, between these two extremes, the earlier concerns are gone, future concerns are far away, and the pregnant woman proudly exhibits her enlarged abdomen temporarily disregarding all else (*dreadful discovery*, breasts, beauty, etc.). The penis-testicle, for once, looks ridiculously small compared to the enlarged abdomen, and the woman is no longer afraid of the *dreadful discovery*. I oppose here the size of the penis and the pregnant abdomen, but the truth is that pregnancy in a woman is far more significant.

Pregnancy is the best positive exhibition, especially when it starts to show through clothing. One has to add the fact that it is holding a fetus, a future neonate, a future human being. It is my personal observation that, no matter under what condition pregnancy occurs, women in some way are proud to exhibit the protruding abdomen. They may hide it behind clothing, but they really do exhibit it. Some recent women's magazines exhibited the fully naked pregnant abdomen. In classes preparing for natural childbirth, husbands accompany their wives to assist in the various exercises. Some husbands are asked to wear elastic bras and an elastic pregnant abdomen to better participate in these exercises. The few husbands I have interviewed in this matter describe the feeling of being pregnant as wonderful.

Finally, a woman will use clothing as a mode of sexual attraction. It is difficult to determine exactly when, for what reason, and which hominid started to wear clothing. Three factors appear to have played a role, and it is not known which one appeared first, whether one factor determined another, or whether they occurred independently.

The first factor is the progressive loss of hair over most of the body. The presence of hair creates a buffer zone between the body and the outside air which protects the body against temperature variations. After the loss of hair, covering the body with some clothing helped establish a zone of air between skin and clothing, and attenuated sudden variations in temperature, thus protecting the body. This was more important at night when bare skin exposed to varying temperatures became very sensitive, thus preventing a deep sleep.

The second factor is migration to the North, which occurred at various periods associated with regression of the Ice Age. The hominids moved to new countries such as North Africa, Asia, and Europe that involved seasonal variations. In the beginning, therefore, the hominids covered their bodies with animal skins to protect themselves against the rigorous climates, especially the winter season.

The third reason has been outlined previously. With the *dreadful discovery* and the need to hide the vaginal and anal orifices, clothing became very useful. Henry, fearful of losing his genital organs (fear of castration), followed Henrietta. Starting with this need to cover the perigenital and perianal areas, Henrietta's need for clothing became greater and greater. She quickly learned that clothing is a potential for sexual attraction. Clothing will be used to reinforce the sexual attraction of each part of the female body (covering of head, different bras, bare legs, etc.).

I will not discuss the recent discovery of the G zone in the female genital area. Unless proved otherwise, I consider that the G zone, assuming it exists, needs further investigation.

Women know too well that all sexual intercourse (penetration, thrusts, ejaculation) is alike. Therefore, it is not in the sexual act itself that a woman will make herself specifically different from any other woman, it is in everything else (anteriorization, superiorization to the breast level, facialization, generalization, and inference of visual, tactile, auditory senses, charm, and beauty) that each woman will be different. On the other hand, the animal sexual act is nothing more than a mechanical one, invariable, repetitious, and stereotyped; sexuality in humans is none of these. Therefore, the more repetitive and stereotyped human sex is, the less it has evolved from the animal level, and the more it is animal-like. The more unique, diversified, and mysterious each sexual act is the more it has absorbed the spirit of evolution.

The human sexual act can be divided into two components, biological and human.

1. The biological part is biochemical and hormonal and is the basic component of all sexual acts. It is similar in all sexual acts and essentially no different from the animal act; it consists of penetration of the male organ in the vaginal cavity, successive thrusts, and ejaculation. It is purely mechanical and involves only the genital organs.

2. The human part, composed of additional elements acquired during anteriorization, superiorization, facialization, generalization, individualization, appearance of emotions, etc., appeared during the hominid evolution. This human part takes a unique aspect for each human individual and for each specific sexual act, according to Kierkegaard (1944). Ideally, no two sexual acts of the same individual are alike. When sexual activity in humans tends to be repetitive or stereotyped the human is bored or repulsed and will look for uniqueness in another sexual partner.

Seventh Reaction, Concept of Beauty

Today, beauty essentially is considered facial beauty. To define it is difficult. Beauty magazines and plastic surgeons mention high cheek bones, wide eyes, and a wide smile showing small, well-arranged teeth. Since nothing in a beautiful face should be prominent, a large nose and a big chin are considered unattractive. Beauty is also extended to a high and large (but not too large) bosom, flat stomach, high tight buttocks, and long sleek legs. Due to cultural variations some women may be physically attractive in some cultures and not in others; one example is obesity. In one extreme, some cultures consider obesity so attractive that women will stuff themselves with food all day long to reach a heavy weight. In the other extreme, some cultures consider slimness so attractive that women will starve to the point of compromising their health and reach a state of anorexia nervosa.

A woman's face always has been to some extent a commodity, and most women consider it their greatest asset. When too much emphasis is placed on visual appearance, it can become counterproductive. When a beautiful person talks stupidly, beauty is gone. The reverse is also true. As a marriage counsellor, I always irritate a patient when I ask him or her to describe in detail the physical appearance of the sexual partner *after* they have finished their sexual act. The Latin saying "Post coitum animal triste" (the animal is sad after copulation) should really be "Post coitum Homo triste." When a relationship is extended in time, beauty is reinforced or diminished with the presence or absence of intelligence, gentleness, consideration, passion, charm, wit, and grace. An error commonly made by many women and men is to consider physical beauty as the only aspect of attraction and to rely continuously and solely on it. Aristotle stated, "Beauty is a far greater recommendation than any letter of introduction," but we always forget that it is only a letter of introduction and nothing else.

If one watches the Miss America or Miss Universe pageants, it immediately becomes apparent how difficult it is to define beauty. There almost never is a unanimous decision in the selection of Miss America or Miss Universe. The elected judges in these contests, although the most knowledgeable in the field, almost never agree on what should be considered when choosing the most beautiful woman. They consider how the candidates look when serious, laughing, smiling; when standing, turning, walking; how they are dressed or undressed. In my opinion, they consider the part of her anatomy she shows, hides, or flaunts; how she hides or exhibits her breasts; what she does with her hair and makeup; in other words, how she artistically transforms her entire anatomical appearance.

Without going any further, we now recognize Henrietta's trials and tribulations during the hominid evolution. We can, therefore, define female beauty, at least primitively, as a mechanism for sexual stimulation. This primitive definition is certainly too narrow to satisfy our present concept of feminine beauty. However, it would be wrong to deny the primitive motivation that led to the search for female beauty. Originally, the search for female beauty was triggered by a sexual attraction that was anteriorized and depheromonized. Today, because it would be considered repulsive, referees in a beauty pageant would never ask the contestants to expose their genital area and rate them according to odor. Yet, if a beauty pageant were held by the early hominids before the sexual evolution started, that is the way it would have proceeded. If male cats or dogs were in charge of a cat or dog show or beauty pageant, this would be their method of selection. However, the present concept of beauty is not defined solely by its history as previously described, although there are numerous similarities between the past and present concept of female beauty.

There are three points I want to make. The first point is that the concept of female beauty is very personal to each individual. While there is general agreement that few individual women are very beautiful, the majority of women are thought to be beautiful by some individuals and not so beautiful by others. The second point is that the reason beauty cannot be accurately defined is because there is still an ongoing search for it. It appears that Henrietta has not completed her evolution in beauty. The last point I want to make is one of history in the original search for beauty. The male body, specifically the penis, is essentially made of angles and straight lines. My impression is that primitive female beauty consisted of round and curved shapes. Today, the concept of feminine beauty is certainly more complex but a beautiful female, in face as well as body, still is conceived as rounded and curved as opposed to straight and angular. "Angular" female beauty and "round" male beauty is also encountered, however.

Male beauty contests are not as popular or well accepted as their female counterparts. These male bodies, in spite of the straight and angular shape of the powerful muscles that are displayed, are not very convincing about their masculinity and appear to have a touch of femininity. In my opinion, these male exhibitions are similar to female exhibitions in that they both are hiding the *dreadful discovery*, at least, originally speaking. Erotic art, mostly Indian or Japanese, exhibit male genital organs of a magnified size.

It is wrong to imagine that the nature of physical attractiveness in women is a decision made by men. Presently it may be so, but originally and *primarily* the features determining physical attractiveness were decided by women, and they still decide among themselves what these features should be: height, weight, body proportions, physical features, the different body parts to be considered attractive, makeup, clothing, etc. They then present it to the males who *secondarily* accept these features as criteria for beauty. Women are more competitive *with* each other, how they appear *to* each other, and comment *on* each other in terms of physical appearance, rather than being interested in men's opinions on those matters. Beauty, originally meant to sexually stimulate men, has now become tremendously competitive between women.

What has been said about breasts can be applied to legs, facial traits, eye makeup, hair style, etc. To my knowledge, a woman never asks for a man's advice about what type of bra to buy, stockings to wear, or makeup, hairstyle, clothing, etc. Men would know little about these items and would not be interested. Men are just asked to respond to all these sexual stimulations.

The takeover of physical beauty as an almost female characteristic is purely cultural and not genetic. There also are cultures where males are just as interested in their physical appearance as females. This point will be discussed later.

Eighth Reaction, Sex as a Service to Man

Why, in every human society, do men pursue women for sex, rather than the opposite? Why do men, at least on first approach, solicit sex from women? Why is sex a favor or service granted to a man by a woman? This is not always true, but it usually appears that way. The answer is simple; it is a question of supply and demand. The sexual urge (estrus) is weak or nonexistent in women, while it is constant in men. Biologically speaking, men need and request sex more often than women. Having questioned hundreds of my female patients and having kept a record of their answers, I concluded that on one extreme female orgasm is almost never experienced more than three times a week. On the other extreme, some women never experience it. Reports by Masters and Johnson are similar. This does not mean that human females do not enjoy sex. Some enjoy sex without experiencing orgasm. This non-orgasmic enjoyment comes from the satisfaction of producing orgasm in the male partner and this sexual enjoyment in a loved partner is a satisfaction in itself. The sensations themselves are also enjoyable. In fact, many females confuse this enjoyment with orgasm itself, but this does not mean that women do not enjoy sex.

The fundamental difference in sex between males and females is that the former is very quickly aroused while the latter is slowly, if at all aroused. I often have asked my patients who is the first to request sex: in ninety percent of the cases, it is a male request, and this is so even among loving couples who communicate freely and who are not inhibited sexually. While women can easily have sex without orgasm, men have sex almost always with orgasm. Because a woman does not frequently achieve orgasm, she can easily control the supply of sex to men. Women quickly took advantage of the situation and made it a service available to men.

This service is becoming more and more important in our culture, and could be classified as the eighth major reaction. From a social point of view, this is the greatest aspect of the sexual revolution. It has become a precious commodity, with access made difficult by women, since they make men pay dearly for it, and women even sell it.

Men court, woo, proposition, seduce, and give gifts in exchange for sex. They also use prostitutes, and rape. Women have something men want; they can provide a favor for one man and withhold from another. In allowing sex, the woman claims to give something for which she receives something in exchange. Henrietta controls the initiation of sexual stimulation and, therefore, has a tendency to saturate Henry with sexual stimulants. She is the one to grant sex, to grant a favor. Courting in the human species is unique in that the female delays sex as long as possible to increase its value to men. If sex is granted too quickly by a woman, it loses much of its value.

Sex has become a service rendered to men by women because they see males having orgasm and most often females do not. Sexual climax in women is much less frequent than in men and much more complicated to reach, since it needs more extensive preparation. It has been acquired during the hominid evolution, and it is only a secondary nature. Therefore, the exuberance, excitement, and precipitation of men rushing into copulation is unknown to women, and they always marvel at the magical effect they produce in men. If we could imagine a world where women quickly and easily reached orgasm but men did not, the reverse would be the rule and sex would be a favor from men to women. In fact, the man who spends the time necessary to bring a woman to sexual climax renders a service to the woman. This will always will be appreciated and he will be courted.

To understand the female favor, one has to consider male activity and female passivity during copulation. Male activity is strenuous and ejaculation comes as a relief; female activity is more restrained and orgasm is easily concealed (as well as easily simulated). All this contributes to copulation being a male rather than a female performance with the male the receiver and the female the giver. This ends in a contradictory situation where the male is an active receiver and the female a passive giver.

If, from birth on, men hid their sexual organs and made them a complete mystery as women do, they would be more attractive to women rather than vice-versa because, at last, they would really have something that could be attractive.

There is another, and in my opinion, more important reason for sex being a favor from woman to man. We have repeatedly seen that sex has evolved further from its animal origins in the human female than in the male. Specifically, the human quality of sex is more obvious in the female than male. A woman becomes more representative of the third dimension of sex than a man, and more symbolic of ideal sex than a man. Therefore, the woman really is giving to man at least the possibility of approaching this third dimension. She is, potentially at least a spiritual giver. I will show later that, as love enters the relationship and emotions are shared, the concept of who gives and who receives becomes attenuated.

Ninth Reaction, Henry's Side

So far I have stressed Henrietta's reactions, but one should not think that Henry has not evolved. He reacted to what he thought he had while Henrietta reacted to what she thought she did not have. If Henrietta did not have any visible genital organs, then he, Henry, also could not have them by losing them. At the origin of the *dreadful discovery*, each and every time Henry had a tendency to proudly exhibit his organs, he also had a tendency to hide them because of fear of loss.

Henry also had some superiorization to the breasts with massive pectoralis muscles in a quadrangular and angular shape. However, this reinforces physical strength and even symbolizes aggression.

Facialization and the concept of beauty in men should never be overlooked, although on first appearance it may appear less intense than in Henrietta. I have no idea when in history man shaved for the first time. I imagine that the initial motivation was to uncover and discover a face behind the beard, although the presence of a beard can be used for "hiding." In public toilet facilities, I have sometimes seen young men spending an unbelievable amount of time studying their facial expressions and combing their hair. One can observe today a revival of long hair and hair styling in men.

Generalization to the whole male body, although more discreet, is certainly as intense as in women. Men, in a different way, also want to be seductive and irresistible and not one of them will overlook the opportunity to look handsome, although in the concept of beauty, the male criteria are different. Unisex hair styles and clothing can sometimes be fashionable.

Tenth Reaction, Clashes and Union

Plato narrates that Jupiter, in his anger against the human species, split the primitive man-woman unity into man and woman. I suppose that the sword he used for the split made an irregular cut, siding the penis-testicle to the male side and the breasts to the female side. In this curse, Jupiter declared that man and woman will eternally try to reestablish the original union, but will always fail, like in sex for instance. It is hard to decide who took the best "cut" in this split. Apparently, man took physical strength, outside activity, and supposedly more "thinking". Women took inside activity, raising children, more feeling, but always felt short-changed because of male abuse.

Men and women, starting with Henry and Henrietta, have always coexisted. The previously-mentioned reactions should in no way be taken as derogatory or degrading to the human condition. These reactions just represent intermediary steps between the animal and human condition, and symbolize the reality of the world from which consciousness was arising in Henrietta and Henry. Human background is animalistic, and the primary reaction to facing other human creatures is not always glamorous. This does not mean that humans are frozen at these reactionary levels. These reactions represent what humans have to deal with to move forward. Henry and Henrietta, in order to move forward, have to confront each other, evaluate, accept, and integrate what they really are. Henry and Henrietta may appear to play tricks on each other, but they really are dealing with each other within the limits of their capabilities and weaknesses. My conviction, also voiced by many others (Robinson, 1972; Schultz, 1969; Teilhard de Chardin, 1957), is that human evolution has not ended but is still ongoing. It may be ended from the point of view of physical anthropology, but human evolution is far from being terminated. The human species is presently at a stage where men and women are pointing fingers at each other, notwithstanding the fact that up to the present time women have been abused by men and are now counteracting.

These reactions should in no way be considered negative,

spiteful, revengeful, or resulting in frustrations. I do not even consider them compensatory. They are just ordinary human reactions and necessary behavior to deal with gender confrontations. These reactions may generate individual clashes but eventually they are preparatory to gender union. Only when these differences are integrated can the union be possible. Only when the human female has gone to the "roots" of what she thinks she is missing, only when the human male has pondered on what he believes is additional in him, can the sexual differences become fully assimilated and used as a bond for union instead of producing friction. No love relationship in the human species is possible if either the male or the female feels "superior" or "inferior."

Presently, women are searching for orientation in every field including sexuality. Female sexuality so far has been determined by human detachment from the animal condition, by prehistory with inherited taboos and rituals, by past and present history. Let us recognize that female sexuality has always been tempered by male decisions. Women as well as men have trouble shaking up these traditions. Although one has to consider the past, the solution is not to be bound by it.

On their side too, men are searching for a new orientation regarding women, especially in the sexual field. Men are well aware of the fact that they have always taken advantage of women, this being done with tremendous variation through history and in different cultures. Most men and women are concerned about reestablishing equality and some are concerned about any backlash.

Men may have the impression of "superiority" but in the long run, it is not so advantageous because emotionally speaking, they are often weaker. Women can take being alone much more easily than men. It has always been my impression that men need female companionship more frequently than the reverse. Over the years, I have questioned the sleeping habits of about one hundred patients who used to share a bed with their husbands and subsequently used separate bedrooms or at least separate beds. Almost uniformly it was at the wife's request and almost never the husband's.

Children mean much more to the mother than to the father, although the latter's love is not questioned. It is just that children fulfill the mother's life more intensely. I will never forget the case of a patient who had nine children over twenty years of a very unhappy marriage. She always stressed to me that when she was hugging one of her children she forgot everything else.

Men are always in need of physical contact with women and suffer more when deprived of opposite sex companionship. Of about fifty couples from my practice that divorced, practically all of the husbands remarried rather quickly, while only about 60% of the wives have remarried so far.

In the long run, everyone loses in a separation. The advantages of a workable union, in spite of many shortcomings, are always far better than the conviction of being able to make it alone.

EMERGENCE OF FEELING AND EMOTION, LOVE AND HATE

Evolution of Feeling and Thinking

For our purposes, a feeling is an inner sensation of pleasure or pain. It is different from physical pleasure or pain, which has an external cause. For instance, a physical pleasure can be derived from eating, drinking, etc., and a physical pain arises from trauma to an organ. On the other hand, a feeling does not necessarily arise from a physical cause and essentially it has an inner nonphysical origin. One also can use the term emotion. The main characteristic of an emotion is that it can exist in relation to another individual and can, therefore, characterize the relationship between individuals. Among the various feelings that can exist between individuals are love and hate.

How much love (or hate) exists between animal individuals can only be speculated on because we do not know how to evaluate their inner feelings. We have no way to verify or study the existence of emotions within the animal individual because animals do not communicate and because there is little expression or evidence of it between individuals. In the jungle, there is an absolute rule that every individual struggles for its own survival. There are no documented cases of an animal sacrificing itself for another. Even the relationship between animals and humans has to be reconsidered in light of conditioned reflexes, although some degree of emotion can be elicited. There are no instances in animal behavior where the emotional maturity commonly seen in humans is seen in animals. All evidence leads to the fact that emotional and intellectual maturity go hand in hand, so there is good reason to believe that emotional maturity in animals is no different from their intellectual maturity (De Unamuno, 1954).

The existence of feelings between individuals supposes the possibility of exchanging feelings or at least perceiving emotions in another individual. Humans have the ability to exchange feelings and perceive the feelings of others through language and using sight, hearing, and touch since these last three senses allow a rich perception of information to go to the brain for interpretation. Animals, however, have no language and cannot receive this information through the three senses, and because of lack of encephalization cannot interpret this information. Therefore, by establishing a parallel between intellectual and emotional evolution, the hominid evolution started with an individual whose limited emotional capability was similar to the present chimpanzee and finished with an emotional repertoire similar to *Homo sapiens.*

Since progressive brain enlargement is considered to be parallel to increased intelligence, it is accepted that there was a progressive increase in intelligence during the hominid evolution. The process began with Lucy, who had a brain size similar to a chimpanzee and ended with the present *Homo sapiens*. The word sapiens (intelligent, wise) means that the present human species has the highest I.Q. among all living creatures on earth. Nobody is denying the value of intelligence in *Homo sapiens* and the role it played during the hominid evolution. Had encephalization not occurred during the hominid evolution, it is likely that the end result would have been some kind of primate not much different, intellectually, from any nonhuman primate. A mistake commonly made, however, is to think that the development of the human brain resulted solely in more intelligence (thinking). This is far from the complete truth since *Homo sapiens* is superior to animals not only in terms of thinking but also in emotions (De Unamuno, 1954). As the brain grew, intelligence and emotional reactions also evolved and grew. Classically, emotions are considered to have the greatest opportunity for expression in sexual activity (Masters and Johnson, 1966). As sexuality was becoming more complex, so were the emotional aspects associated with it.

Some investigators have accepted the fact that there has been no evolution in the field of feelings and emotions (love, attachment, affection, and bonding between individuals) and that extant animals have the same feelings as humans. Only a few would accept that there has been a radical difference and evolution from animals to humans in the domain of feeling. In many circles, the emotional aspect of sex is considered the same in animals as in humans. Some animal lovers have even stated that, in terms of emotions and feelings not only have we failed to progress but we have markedly regressed.

I will show that feelings and emotions differ when comparing animals and humans, and feelings and emotions evolved during the hominid evolution. How this emotional transformation from an animal to present human level occurred will be studied in this section.

Anatomical Basis for Emotion and Individualization

There is another important difference between humans and animals in terms of brain activity. Compared to the human neonate, what the animal neonate learns from birth on is limited, and it occurs mostly by copying the mother and others. On the contrary, the difference in intelligence and emotions between the human neonate and the human adult is tremendous. The human neonate is born with very little to face life on its own, and this remains so for a long time. Almost all animal neonates are born with a kind of genetic knowledge system that is either immediately available (precocial neonates) or very quickly available (altricial neonates). Only the human neonate comes into the world with practically nothing and has to learn everything. I find it astonishing that a newborn calf can use sight, hearing, touch, smell, and taste as soon as it is born. It seems as if this animal neonate was part of the outside world well before it was born and that it can function in an easily recognized and familiar home as soon as it is born. When the human neonate is born, it is a rather frightening experience since it knows nothing about what it sees, hears, touches, or even smells or tastes, and all it can do is cry. The human neonate has to learn to recognize objects and faces, to interpret what it sees and hears, learn language and the meaning of touching, and so on.

The anatomical interpretation of this learning process is very simple. After human birth, synapses, axons, and dendritic formations (extensions of the brain cells or neurons) in the brain are continuously formed. The process is very active during infancy, childhood, and adolescence, and can be continued throughout one's entire life. Of course, this progressive and continuous brain formation after birth is much more intense in humans than in any other animal. The human brain is not only the largest at birth (relatively speaking), but it is the one that enlarges the most after birth. Modern research has demonstrated that synapses (connections between brain cells) continue to form well into adulthood. More important, synapse formation is different for each individual since the input that is received varies depending on the individual. A tremendous amount of interconnections between the different zones of the brain, and formation of these interconnections mostly after birth, are the essential traits differentiating the human from the animal brain.

The conclusion from the above is that human cerebral function (intellectual as well as emotional) is in great part acquired after birth and continues to be learned and formed during life. This is how we differ from animals who acquire much less after birth. This leads to a new definition of the human condition as opposed to the animal condition. Anatomical characteristics are what differentiate various animal species, but within a particular species there are limited differences between individuals. In the human condition, it is in their intellect and emotions that individuals completely differ from each other. The more a human individual grows independently and distinctly from another individual, the more that individual will be different from others. In large part, their brain formation is postnatal and acquired, while in animals it is prenatal and genetic. It is in the field of sex that these individual differences are most clearly expressed. The main difference between animal and human sex is that animal sexuality is stereotyped, and human sex is individualized, intellectually as well as emotionally. The reason is because it is formed more after than before birth.

The areas of the cortical brain associated with the functions of feeling enlarged during the hominid evolution as much as the other areas of the cerebral cortex associated with intelligence. It is now proposed that half the brain is concerned with feeling and the other half with thinking, although this is only partially accurate and is still debated. If one accepts the concept that half the human brain deals with thinking and intellectual operations and the other half with feelings and emotions, then, from an evolutionary point of view, the second half of the human brain is just as important as the intellectual half.

Thinking and feeling are the two main functions of human mental activity, which separate and differentiate us from all animals including higher primates (chimpanzee, gorilla, orangutan). This is anatomically supported by a more developed brain. The previous generation of researchers did not hesitate to use the term *subhuman* primate to designate animals such as the chimpanzee, gorilla, orangutan. Today, out of respect for the animal condition, the preferred term is nonhuman primate. I have no quarrel with such a change in terminology but, to avoid confusion, this change has to be considered in terms of what motivated it. The statement that humans are not only *different* from but also *superior* to animals in terms of thinking and feeling may seem naive, except that more and more it is being challenged and even denied, especially concerning feeling. I agree that I have magnified the differences to make my point, but basically humans have markedly evolved from the animal condition intellectually as well as emotionally, for better or worse.

Loss of Aggression and Appearance of Feeling

By nature, animal sex is aggressive. Aggression in animals is not an emotion as we conceive it in humans, it is a modus operandi. It is a mistake to imagine that two animals who are having coitus or fighting are expressing love or hate, the way two humans would do under similar circumstances. Actually, neither love nor hate as conceived by humans exists in animals. Animals are aggressive to each other not because they hate each other (the ways humans would do), but only to get their way, that is, establish hierarchy, defend territory, get food, and take a sexual partner.

The fundamental difference in sexual behavior between animals and humans is the decreased amount of aggression in humans, although it does occasionally surface. In animals it is more frequent. Because the human sexual act supposes full cooperation on the part of the female, a change in behavior (disappearance of aggression) became necessary. How and why sexual aggression disappeared will now be described in detail. Essentially, if humans were going to survive, abandonment of aggression and the birth of love and emotional attachment were desperately needed in two specific fields: mother-infant bonding and sexuality. There are many reasons why the revolution was necessary in these two fields.

First, the biochemical system that determines the animal mother-infant bonding was collapsing. Placentophagy (consumption of the placenta by the mother immediately after birth) and amniotic fluid licking (commonly practiced by the mother after birth) were disappearing, and prolactin and lactation were becoming less effective as a mode of bonding in the prehuman and human species. Olfaction and taste, which were powerful stimuli in the biochemical system, were diminishing and being replaced by new neocortical functions such as sight, hearing, and touch. The newborn was becoming more altricial, that is completely helpless at birth and for a long time after because, compared to animals, the human neonate is considered too premature to be on its own. Women were becoming more helpless during pregnancy, labor, and delivery and needed support and protection as infant caregivers. Some new neocortical function combining emotional and intellectual requirements of mother-infant and individual bonding had to appear for the sake of survival of the hominid species.

Second, a new approach to sexuality was necessary similar to the new approach in mother-infant bonding. An emotional bonding to replace the departing biological bonding, a new bonding strong enough to motivate getting involved in this complex human sexuality was a primary requirement for the survival of the human species.

In the emotional revolution, Henrietta had more at stake than Henry in the early stages because of the loss of pheromones and anteriorization of sex, while Henry had less to change. It was only later that Henry caught up with Henrietta. Without appearing prejudiced against Henry, I would venture to say that he carried traces of sexual aggression for a long time and may still carry them. The sexual prehistory and history of mankind has a long list of physical and mental abuses of women by men.

Henrietta played an important role in this emotional revolution and in the attenuation of male aggression, sexual or otherwise, at least in the beginning. Henrietta and Henry badly needed to control their aggression and the biting in the ventral-ventral sexual posture. Disappearance of the canine teeth attests to this fact. Henrietta's new weapon will be to use feeling and charm, which she will promote through the newly-expanded senses of sight, touch, hearing, and vocalization. Henry will use intellect, which, as I will show later, was expressed more frequently. But really, Henry and Henrietta are both using feeling and intellect in overcoming physical strength and brutality and the hierarchy based on physical strength. Individual and emotional potential now come into play and are used more than aggression in measuring human value. I have previously shown (Chapter IX) the role played by the dynamic use of vision, hearing, touch, vocalization, and body movements, etc. in *intellectual* development. I will now show their role in *emotional* development, which is definitely the most important step in the birth of human sexuality.

Persistence of Biological Sex

The sexual hormones (androsterone in males, estrogen in females) are present in animals as in humans. Human sex has never lost its biological and mechanical aspects, and in this respect it is like sex in any other mammalian species: penetration, ejaculation, withdrawal. The similarity ends there, however. The animal sexual act consists of nothing else while the human act has the potential for much more. Nothing precedes or follows the animal act, while the human act is preceded by, associated with, and followed by feelings, emotions, and actions.

Human sex is not biological or mechanical as in animals and, although in some respects it may appear so, the individual in that case is just turning away from the human condition represented by three million years of evolution. Animal sex is more biological and mechanical than emotional, and whenever humans attempt to practice it the animal way (which I do not think is entirely possible although some pretend it is), they expose themselves to the repression of emotions, which is always worse than the expression of emotions. As sight, hearing, and touch played an important role in the rise of human sexuality, so will they play a role in the rise of human emotions.

Role of Sight in the Rise of Emotions

As previously mentioned, one of the fundamental differences between animal and human sexuality is that the male animal looks only at the back of the female and "all backs look alike." In humans, the sexual act is face-to-face and the partners look at each other. This is called eye contact and is even prolonged eye contact. Eye contact in animals, even in higher primates, is very short because they do not identify and communicate with each other by sight. Humans, however, use eye contact continuously. It is almost impossible for two persons to identify and communicate with each other without eye contact. Relationship or communication without any eye contact is a form of severe mental disturbance. In a face-to-face sexual act, one is automatically led to look at the whole face of the sexual partner.

I can only guess what primitive Henry and Henrietta thought while looking at each other eye to eye in a sexual act. They could no longer hide how they feel about themselves, each other, or what they were doing to themselves and each other. The truth surfaced because Henry and Henrietta had all the time in the world (three million years) to figure it out. Pain, fear, and terror took unmistakable facial expressions. Henry easily read the effect of his aggression and brutality on Henrietta's face. He saw that he was not wanted and was feared, disliked, hated. Reciprocally, Henrietta learned how to express her emotions. Her face expressed the corresponding feelings, and Henry saw acceptance and desire. Henry learned more and more that certain behaviors, sexual or otherwise, were associated with fear and repulsion and others with desire and attraction. A mutual flow of emotions was established. Later on, Henry also might have perceived a third type of impression in Henrietta, a neutral impression, neither pain nor enjoyment but just boredom, and Henrietta learns how to express it. In summary, Henry saw that he could elicit pain, pleasure, or boredom, all of which he could read on Henrietta's face. This flow of positive and negative emotions started from Henry to Henrietta as well as from Henrietta to Henry. Love and hate were born.

Henrietta was dynamic because she quickly learned how to elicit the different emotional states in Henry.

Use of Touch in the Rise of Emotions

The use of free hands tremendously modifies the characteristics of the body embrace during the human sexual act. The hands and arms (now free in the ventral-ventral posture) can be used to push away or to express resentment on Henrietta's part; Henry can use his hands aggressively to choke Henrietta to impose himself; or the arms and hands can be used to reinforce the body embrace. Henrietta can also express indifference to the sexual act by lying passively with her hands on each side of her body.

What can be done with free hands in terms of human sexual activity can be tremendous. The new function of touching, with hands and/or any other part of the body for that matter, creates sexual stimulation. Feelings and emotions can be expressed through touch with the two partners learning to recognize what produces pleasure, pain, or indifference. Touching between men and women takes on special significance and easily can be interpreted as sexual in nature, leading to sexual arousal. In fact, in addition to sight, sound, smell), it is the most common sexual stimulation in humans and it is entirely under conscious control. I consider sexual stimulation by touching the best substitute for the departing pheromones.

Hearing and Vocalization in the Development of Emotions

Vocalization associated with any visual or tactile action now takes on special meaning. Vocal expressions of hate, fear, and pain, or pleasure and satisfaction will now be recognized by Henry and Henrietta.

For the first time, vocalization is comprised not only in shouts or threats (negative expressions) but also as expressions of satisfaction, fulfillment, happiness, and ecstasy. Most likely, a sexual or love language was the first communication between Henry and Henrietta. At its origin, language must have been musical tones of happiness or sadness, associated with sex.

The Development of Emotions and Feelings

Thus, a new type of sexual activity that includes emotions is being built progressively. Feelings and emotions are born during the sexual act, are easily transmitted from one sexual partner to the other, and are expanded to other aspects of hominid life. The smile, specific to the human species, appeared somewhere during the hominid evolution. The smile, which is basically the expression of a meaningful relationship between two individuals, expanded into laughter. It most likely expressed relief at the disappearance of aggression. It expressed the enjoyment of sex. Unfortunately, pain, fear, and repulsion are also visible and are expressed with tears and crying. Feelings and emotions born during the sexual act and expressed by smiling, laughing, and crying are still the most powerful means of emotional relationships used by women in dealing with men, and vice versa. They easily replaced the departed pheromones and became just as powerful.

The Concept of Individualization

Monogamy has been claimed in isolated primates. From these isolated observations, the concept that some animals are coupled for life became accepted. Many of these observations are superficial and the conclusions are certainly hasty. The formation of sexual grouping in animals (including higher primates) has to do with many known and unknown factors. For practical purposes, a permanent one-to-one relationship is specific to the human species and started sometime during the hominid evolution. The absolute prerequisite is identification and individualization of the sexual partner, which barely exists in animals. In the beginning of the hominid evolution when Henrietta first stood up, there was little individualization. Henrietta and Henry progressively learned to differentiate and individualize sexual partners depending on the feelings and emotions elicited by each one. Evidently, the amount and quality of feelings and emotions elicited during the sexual act depended on the sexual partner. Not all sexual acts were the same anymore because not all sexual partners were alike and, therefore, the emotions elicited by different sexual acts were not alike. A tremendous step forward had now been made in the human condition; namely, recognition and individualization of the sexual partner. This represents the crucial turning point in sexual and human evolution in general.

By now, hominids learned to establish differences between individuals. Henrietta and Henry started to recognize the different emotions they experienced during the sexual act and associated them with an individual sexual partner because they had different visual, tactile, and auditory characteristics. Memory capacity grew because, as sexual activity expanded from seasonal to year-round, Henrietta and Henry could remember the facial features of the individual who produced the specific feelings and emotions. Even between sexual acts, Henrietta and Henry could look forward to being with or avoid a specific individual. Human individualization, which started with sexuality, is one of the crucial landmarks in human evolution.

It is not important whether individualization started as a general behavior with sexual individualization a consequence of it or began in the sexual act and expanded to general behavior. Either way, sexuality played a crucial role in the formation of individuality and this is the point I am stressing. Henrietta played a more important role than Henry in this individualization process in the beginning because, as previously noted, she had more at stake. Eventually, men also individualized their sexual partners.

Identification by facial and general body features outside of sexual activity followed. The body's visual appearance plays a limited role in animals and their facial appearance even less. Facial identification is certainly sexual in origin, or at least sex contributed to it. It is always exclusive to humans.

Sexual Privacy

Sexual privacy, in addition to adding to the mystery, reinforces individualization. It allows for more intimacy (also fantasy) and becomes a more unique experience. In extreme cases, sex in darkness helps partners escape the eye-to-eye contact which might reveal an intimacy so deep that most people cannot tolerate it. At the moment of sexual climax, not too many people look at each other, especially when they reach orgasm simultaneously.

Emergence of Sexual Orgasm

I have shown that there is no orgasm in the animal world, including present higher primates, although some authors have described rudimentary physiological elements resembling an orgasm in primates. Assuming it is a real orgasm, it is very primitive. The real sexual climax is specific to the human species and, therefore, appeared during the hominid evolution. I will now describe how it might have originated.

It is possible that in the beginning, when Henry and Henrietta started to practice ventral-ventral sex, the female became aware of the male orgasm regardless of how primitive it might have been. As the male orgasm evolved and became more visible, the female desired to build her own climax in order to reciprocate. Duplication of, and participation in, the male climax might have started this way for Henrietta. The production of something similar to male orgasm is one thing, real arousal of female orgasm is another. Either one or both (or neither) may occur. The problems (arousal of female orgasm and participation in the male orgasm) are still contemporary ones. Human female orgasm is not the rule in each sexual act the way male orgasm is, since male ejaculation usually means orgasm.

In our present human culture, it can be irritating for a woman to watch the male easily reach orgasm, while it takes more time for her. The inability to copy the male can lead to female frustration, but this does not mean that a woman who does not reach orgasm is always frustrated. The process of individualization is so intense in women that the sexual act with the loved one or preferred partner is a satisfaction in itself, even if not associated with orgasm. Absence of female orgasm does not prevent coitus, and in this case female response is indifference and passivity. In fact, the most important difference between male and female sexuality is that the woman can be completely passive and yet perform sexually, while the male has to be fully active. A frigid female (who does not reach climax or achieve any pleasure in sex) can get involved in sex (even become pregnant) while an impotent male (who does not achieve erection) cannot. There is no such thing as an impotent female or a frigid male.

Sexual orgasm in humans is a result of sexual stimulation built up by the two partners during the sexual act itself. The appearance of sexual orgasm was necessary anyway as a substitute and replacement for the departed pheromones and weakened estrus. Natural selection was for those who were capable of experiencing sexual orgasm and of eliciting it in the partner. In view of the complexity of human sex and the marked weakening of the biological drive, *nothing* would motivate humans to perform sex if it was not for the emergence of orgasm. Sexual orgasm became necessary for the survival of the human species.

Orgasm is a reciprocal growth in sexual maturity between man and woman. Freud says the clitoris (penis like) is primary and the vagina is secondary. He considers, therefore, that clitoral stimulation is masculine and active while vaginal stimulation (coitus) is passive and feminine. He also suggests that clitoral stimulation occurs during childhood and considers that clitoris and phallus are alike. I question this approach. In relation to its size, the penis is not that active in childhood, and I believe that childhood masturbation of the clitoris is exaggerated by Freud (1963). I do believe that the role of clitoral manipulation by girls and masturbation by boys has been overplayed by Freudian psychologists.

Frigidity or inability to reach orgasm is considered an abnormality in women; in my opinion, it is a primary normal pattern. Orgasm is a secondary condition acquired after proper maturation. The proof (Helena Deutsch, 1945) is that no form of psychotherapy cures frigidity. Unfortunately, most people (as well as psychologists) think orgasm in women is normal, as orgasm is normal in men. They overlook the fact that ejaculation equals orgasm in men, forgetting there is no ejaculation in women, although it has been suggested that some women have a small fluid expulsion at orgasm. This results in fear, guilt, and resentment. Frigid women need emotional growth and/or a love relationship, not psychotherapy.

Real female orgasm is not general and psychiatrists, psychologists, marriage counsellors, and other specialists who approach the problem objectively know it too well. The other well-known fact is that real female orgasm is acquired slowly and progressively after a prolonged process of intimacy with a specific human being. A third fact is that most women who do not experience orgasm have no idea what sexual orgasm really is. A woman in love will easily confuse and be satisfied with sexual intimacy as an orgasmic experience. Another woman will empathize and participate in the sexual excitement of the loved partner as if it was her own excitement. Many women enjoy the ability of producing such sexual excitement in the male and are satisfied whenever they exert this role without even participating. In addition, my impression is that during sexual intimacy the woman integrates the male sexual orgasm as her own and when the male organ penetrates her deeply, she fantasizes for a brief moment that she has corrected the *horrible discovery*. Finally, many women learn (consciously or not) to simulate an orgasm rather than getting involved in the elaborate procedures that are required to achieve it. The reason women are predisposed to calling almost anything an orgasm is the fear of being tagged as frigid. This would be wrong because frigidity is the lack of interest in sex.

Women who experience real orgasm have their own specific way to describe it. It is composed of repeated spasms of the genital and perigenital area associated with a state of complete relaxation, the desire for continuous penetration by the male even after climax, the feeling of intimacy, an ethereal condition with an impression of elevation. All are unique experiences, not quickly forgotten. Diane Ackerman (1990), the sex poet, described this experience very well. A woman who experiences orgasm never forgets it.

In a parallel but different way, Henry was also progressively building his own orgasm, but his was easier because it was based on the biological ejaculation.

Somehow male and female orgasms have been lumped together under the term sexual climax. I do not believe these should be equated and compared, but assuming this is to be done I have several comments on this matter:

1. Male orgasm is natural and biological or, at least, is combined with the result of biological phenomena such as seminal vesicles and vasa deferentia congestion, and is associated with the brutal expulsion of sperm through the urethra. Since the male urethra is an elongated tube (more so than in the female), the sperm under pressure, rubs extensively against the urethral mucosa, and ejaculation produces the intense sensation of orgasm. To a limited degree, it also can be produced by urethral catheterization (a medical procedure introducing a catheter into the urethral canal), by prostatic massage (medical examination of the prostate which stimulates the proximal segment of the urethra), or simply by strong urination under intense pressure.

Sperm ejaculation under strong pressure is biologically necessary as it ejects the sperm into the cervical canal of the uterus, thus facilitating impregnation. This leads to the fundamental difference between male and female orgasm: male orgasm fulfills a biological function while female orgasm does not. The male orgasm is "natural" (determined by ready-made anatomical structures and specific physiological mechanisms), while the female orgasm, not connected to any specific anatomical structures or physiological function, is much more difficult to define anatomically or physiologically. This is the reason it does not exist in the animal female. The frequent mention that female orgasm relaxes the genital tract thus facilitating sperm ascension is pure fantasy without any scientific basis. Another frequently mentioned statement that women are predisposed to orgasm during ovulation is also without basis, since orgasm can be reached at any time during their cycle, even before puberty, during the menstrual period, during pregnancy, and after menopause.

2. The above description in no way implies that animal and human male orgasm are alike. The human male orgasm is far more elaborate than that observed in animals. It just has animal roots of a biological nature. The human female orgasm has no ani-

mal roots since it is not necessary for procreation and is an entirely human creation.

3. With very few exceptions, the human male almost always reaches orgasm during the sexual act, and this leads men to think they have an advantage over women. However, real female orgasm is much more intense and of a very different nature. I have repeatedly discussed in depth with different men their feelings and impressions before, during, and after orgasm. The responses I have gathered are monotonously repetitious: a desire to eliminate all preparations (foreplay) if they are prolonged; obsession with ejaculating as soon as possible; withdrawal after ejaculation and turning over and falling asleep; need to start over the same day or next day; tendency to look for different sexual partners; obsession in testing new sexual postures, etc..., and I am cutting this list short.

It is not unusual that the human male sexual act, although always orgasmic, is associated with sadness, guilt, and a feeling of isolation especially when frequently repeated.

4. The really orgasmic human female presents a totally different picture. There is no desire to repeat the orgasm the same day or even the same week. I have pondered on this strange finding until one day a patient gave me the answer. Although her orgasm produced feelings of ecstasy for a few hours afterward, there was also a feeling of contentment and well-being for the rest of the day and of satisfactory quietness for a few days afterward. This patient told me that another sexual act too soon after the first one may disturb her state of relaxation. She even added a humorous tip: whenever she had an evening party or social gathering to attend she tried to plan an orgasmic sexual act a few hours before (instead of spending hours in front of a mirror) and, without any special makeup or glamorous clothing, she always appeared at the party with such a fascinating glow that she was noticed by all. She never told her secret to anyone, but I have passed on the tip to many of my patients. This observation on the prolonged effect of the female orgasm has been confirmed to me by many patients. When I ask women to describe their feelings before, during, and after orgasm, I have yet to find two women who give the same description of their orgasmic experience.

5. When comparing male and female orgasms, there is a dis-

advantage on the male side because men cannot have orgasm without ejaculation and the formation of the ejaculate supposes that the secondary sex organs (vasa deferentia, seminal vesicles, prostate) again become filled with sperm. Therefore, the male has to wait for their refilling before another orgasm, and this may take many hours and even a whole day (contrary to animal males in heat who can repeat ejaculates within a few minutes). The female has nothing to ejaculate and can, therefore, repeat copulation as frequently as she wishes.

I will terminate my description of the human female orgasm with an interesting note. I have known some men who have been so fascinated by the depth of the sexual orgasm of their female companions that they have tried to duplicate it by timing their ejaculation with the female orgasm. They just did the reverse of the average woman when she tries to duplicate male orgasm.

In the complex ejaculation-orgasm of the human male the ejaculation is identical to that of animal males, but the orgasmic part is not. There is ejaculation similarity only. On first approach, one is erroneously led to believe that ejaculation is orgasmic in animals and even that human male orgasm is animal in origin.

I conclude that female orgasm is not biological, and not natural, at least originally speaking. Neither is the real male orgasm biological, only the ejaculation part is. Aside from this common definition or goal, the two orgasms are anatomically and physiologically completely different.

Sexual orgasm in the female, as well as in the male, is the greatest gift to humans by the female gender. In fact, I consider orgasm more of a driving force than the departed pheromones because the latter did not have any effect beyond the sexual act itself, while the former has long-lasting effects. Orgasm made human evolution possible. In fact, this is how *Homo* and *Femina sapiens* were born. Henrietta really initiated it and Henry joined eagerly.

Orgasm, Ejaculation, Reproduction

With no more sexual stimulation by the departed pheromones, nothing would lead the male to perform copulation, if it was not for the newly acquired male orgasm. Male orgasm was born and natural selection favored orgasmic males. For this reason, there was no need for female orgasm and her usual passivity was all that was required for copulation/reproduction. Human female orgasm has no place in natural selection and has to find a justification that has nothing to do with reproduction.

The embarrassing conclusion is that from the reproductive and evolutionary points of view, male orgasm is a natural biological necessity but female orgasm was not. This does not mean that the human female does not need any orgasm.

The tragedy and error made in female sexuality is to look at it in male dimensions. The average woman in attempting to achieve orgasm considers it solely in terms of male characteristics. She is "perfectly" orgasmic when she mimics the male orgasm in every respect, including gestures, behavior during climax, duration of orgasm, repetition, and frequency. The most popular articles or books on female sexuality describe this similarity, reinforce the female potential for a male-like orgasm, and even (most comical!) describe the female genital area as possessing all the anatomical elements necessary to reproduce the male orgasm such as clitoris is like a penis, corpus callosum is like crux of the penis, etc. (Sherfey, 1973). One always forgets that a woman does not have sperm to be ejaculated and even if she had anything to ejaculate, it would be through a very short urethral canal anyway. This forced similarity is ridiculous as well as erroneous.

In the human male, ejaculation and orgasm form an indivisible complex necessary for fertilization. This is a crucial difference between men and women; namely, for a sexual act to result in fertilization, male orgasm is mandatory and female orgasm is unnecessary. Yet this difference has barely attracted specialists on human sexuality. The consequences of this bizarre but important observation are numerous. Successful reproduction is what evolution and natural selection is about (Trevathan, 1987). Assuming this is true, and it certainly is, natural selection would favor males with powerful orgasms but would pay no attention to female orgasms. My dreadful impression is that at some point, and even at several points, during the hominid evolution the very existence of the sexual act was on the verge of disappearing or at least threatened to disappear. This may explain the paucity of the hominid populations that preceded modern *Homo sapiens* and even the disappearance of all of them. After all, one may not find an immediately available and suitable replacement for a failing organ or function, and natural selection may take its time to find the proper substitution for the departed pheromones as a sexual stimulant.

The female who thinks that she will reach orgasm by simply getting involved in a sexual act is in for a shock. Practically every female who starts sexual activity abruptly without carefully preparing through trial and error, without having an intimate relationship with a specific partner, or having the proper emotional maturity (and even after preparation) is rather disappointed. Female orgasm is fragile, volatile, not easily reached. Lucky is the woman who finds the proper circumstances and a skillful and patient lover to assist her for an orgasmic sexual life. Unlike men, women do not always reach orgasm at each sexual act and this basic difference between male and female sexuality determines their separate sexual behavior.

Concept of Love

If the first greatest gift of women to the human species is sexual orgasm, their second greatest gift is the emergence of the concept of love. Again, man eagerly followed. Once the emotions of one sexual partner are perceived by the other and both partners perceive and initiate each other's emotions, then the road to love is open. How to define love, at least in the primitive form when it appeared during the hominid evolution, is a much simpler task than to describe it in its complex form in present humans. Primitively, we can define love as a commonly shared feeling with the desire to reexperience it. Each partner has the ability to trigger an emotional feeling in the other and when the feeling is reciprocated, then love follows. It is in the sexual act that love is mostly experienced, at least primitively. It is no surprise then that Henry and Henrietta experienced their first and deepest love in the sexual act. The sequence of events leading to the concept of love was certainly elaborate in hominid sexuality. I have described how emotional feelings (either pleasurable or painful) arose during sexual activity, and it is likely that for a long time during the hominid evolution, Henrietta and Henry were just building up feelings during sexual activity without necessarily relating them to a specific sexual partner. Initially, the concept of individualization barely existed but, once it was well established, it was easy to relate the emotional feelings associated with the sexual act to the specific sexual partner who elicited it.

With the development of memory, the sexual partner could be recognized even outside the sexual activity and a preference for that individual developed, especially when the individual was not aggressive during and between sexual acts. All the elements for the emergence of love are set. The concept of love was not originally reciprocal; far from it. Henrietta was the first to fall in love, although Henry's orgasm gave him positive associations to Henrietta. Only later with patience and skill and helped by the emergence of her beauty and of her smile did she finally communicate, at least partially, her feelings of love to Henry. After much struggling, Henrietta and Henry finally changed their old ways.

The birth of emotions and love (at least the possibility) has slowly replaced aggression as a mode of sexual approach. Using a simple physiological act (the coital act), Henrietta and Henry slowly built up love, a concept on which present human culture rests.

Primary love is between man and woman with all the positive and negative psychological aspects that go along with it. Love extended afterward to other aspects of human life besides sex. Secondarily, love extended to the mother-infant relationship with all the aspects of primary love projected and displaced on it. The mother will react to the infant the way she relates to her sexual partner. I am convinced that the first emotional bonding was not the mother-infant bonding as seen in animals, but between men and women performing sex. Only after Henrietta felt love in sex, only after she established the connection between sex and pregnancy was she able to transfer this love to the product of her love relationship. After all, children who are the most loved are the product of a sexual relationship associated with love. The mother who felt love in sex first was able to transfer it afterward to the child. The human mother could then face the departing biological mother-infant bonding of animals and replace it with the emotional mother-infant bonding of humans. Trevathan (1987) said, "A common assumption, however, is that all these bonds between humans ultimately can be traced to the very first one, that of mother and infant." I disagree with such a statement; a man-woman relationship is the primary bonding. As previously elaborated on, the mother-infant bonding in animals is purely or mostly biological, while the human one is not. Togetherness in animals is limited strictly to the time spent in copulation; otherwise they practically ignore each other or establish hierarchy. Men and women spend much more time together than the duration of copulation, especially when there is a love relationship.

Concept of Hate

It would be wrong to consider that love is the only human feeling originated in the sexual encounter following individualization and the appearance of the female orgasm. Unfortunately, expansion of emotion and feelings was not always positive. The sexual act, which was casual in animals, now took on a meaningful dimension and was very traumatic when not associated with selection and individualization of the sexual partner. Henrietta learned to hate Henry when sexually aggressive and Henry learned to hate Henrietta when sexually noncooperative. As the human female learned to love in sex she also learned to hate in sex. The appearance of hate, which is the irrational desire solely for destruction of another human individual, is born. Hate does not exist in animals for the same reason that love does not. Aggression is just a Darwinian technique to allow for survival of the fittest over the less fit. Once the less fit have been eliminated, the animal has no desire to pursue the destruction of the weakest. Hate is completely different since it has no other purpose but to destroy and harm another individual. It is specific to *Homo sapiens*, the way love is also specific to *Homo sapiens*. Hate can destroy as much, if not more, than survival of the fittest because it is irrational and can reach unbelievable proportions. The hatred and destructive tendencies that humans can express to one another can reach incredible proportions as it occurred during the Nazi holocaust.

Whenever individualization, feelings, emotions, or love are frustrated, the result is jealousy. Jealousy is specific to the human species and is the result of frustration or suspicion that the one-to-one relationship created by sex or otherwise has been broken. Only humans are jealous, although some animals living in contact with humans can also be jealous. Hate, or persistence of jealousy, occurs when jealousy is not relieved. Jealousy started with individualization of the sexual partners and becomes more intense when the one-to-one relationship is increased and maintained beyond copulation.

Sexual Sequence in Humans

Let us now describe some other aspects of the human sexual act that are also different from that of animals. In humans, the process consists of five stages:

1. Visual message. The "look" with special sexual significance and attraction may be difficult to define, but it always sends a clear message. The facial expression takes on a special appearance with a vague persistent smile aimed at the selected partner. Leon Uris (1989) in "The Haj" describes it as the only message a Moslem woman can use when her entire face, except for her eyes, is covered. A single visual message may be enough, but if necessary it can be repeated on several occasions over extended periods.

2. Auditory message. Combined with the visual appearance, it becomes courtship. What to say and not to say, and how to say it, has become an art. This is how the two partners learn to know each other. They decide then whether to pursue or discontinue the relationship. The concept of courtship, at least the way it is conceived in humans, is seen only in humans. Nonhuman mammals display aggression between males competing for females, or against the female to force her to the sexual act. These are the only signs of courtship seen in nonhuman mammals.

3. Tactile relationship. This factor is essential because sexual stimulation in humans is mostly tactile. It is, therefore, the most

crucial aspect of the relationship, the one that determines whether or not there will be a sexual encounter or if it produces a desire on the part of the female for more encounters. This is the part most often neglected by the male partner, either out of ignorance or, simply because he had only the sexual encounter in mind. A male who patiently and gently builds up the female partner to a sexual climax is assured of the female wanting more sexual encounters, assuming he does too. At this stage, sexual inhibitions and proper sexual preparations play a role. Egotism, selfishness, altruism, or concern for others, play a role. This is where the real personality comes in and where the animal or the spirit shows up.

4. Sexual encounter. The sexual act, although often described as biological, does not have to be. It can be short if the male wants to ejaculate quickly or it can be prolonged for the female to reach orgasm. For the male to learn to refrain from a quick ejaculation, which would brutally interrupt the sexual act and to allow the female to reach climax, requires numerous human qualities (love, patient, caring) not always found.

5. The after-sex relationship. In my practice as an obstetrician and gynecologist, I often act as a marriage counselor and sexual advisor. Frequently, I am faced with young people who are sexually active, even promiscuous, and who do not understand why they have a sexual problem. I always elicit their opinion of themselves by asking how they feel about themselves and each other *after* the sexual act. It is mostly in a casual encounter that quickly ends in sex with the two partners never meeting again, or in an act of prostitution, that I cause most embarrassment. Many of them refuse to go further in counselling, but some are willing to go deeper. The feelings of the two partners toward themselves and each other are easily discerned by describing their feelings after the sexual act: hate, boredom, indifference, desire to leave immediately, or on the other hand, mutual ecstasy, enjoyment of togetherness, holding hands, etc. Animals walk away from each other after coitus as if nothing happened or as if they never knew each other. Humans are different in that there is always an after-sex episode.

6. Building up the relationship. Either the relationship builds up progressively with mutual enrichment, understanding, and acceptance or the partners drift apart and separate. In this case they

start over with a new partner. It may become a spiritual relationship or an addiction to repeated relationships. In repeated relationships (promiscuity) only the biological part is repeated. Frequent sexual encounters with different partners and the inability to sustain any of the relationships for a long time is actually a biological regression to the animal level.

The sequence in animals is different and consists of olfactory attraction, fighting competitors, overpowering females, sex of very short duration, and walking away. Animal and human sex belong to two different worlds with the possibility of humans partly regressing to the animal level.

The New Woman in the Sexual Revolution

To say that the female hominid used deceits, tricks, etc., during the hominid evolution would be a tremendous misconception. To think that the emergence of human sexuality was based solely on hiding and flaunting would be a complete misunderstanding of the essential nature of sex. Some of these steps can be seen in women today, but they should be viewed only as available tools used to build human sexuality, to build a relationship based on love, mutual respect, tolerance, and enthusiastic acceptance of the differences in the other. Most intolerances are nothing but repressed aspects of our own personality which are projected and criticized in others. Tolerance of others, specifically of the opposite sex, is therefore the simplest way to accept oneself totally. I repeat, the first gift of the female gender is the gift of human sexuality including sexual orgasm, and the second is the birth and expansion of feelings and of love relationships. Life in general is arduous and difficult for the human species and tremendous motivation is therefore necessary and was provided by women and easily received by men in the form of human sexuality, orgasm, and love. Love was extended to other aspects of human life, mainly mother-infant bonding or rather parent-infant bonding. Without love, at least without the potential for it, humans would not have survived as a species because there would not be enough motivation to go through the struggle of life.

In its goals, if not in its steps, the birth of human sexuality has great positive value. There is a positive aspect even in the *dreadful discovery*, a desire for understanding and acceptance of the difference and to form bonding. My conviction is that bonding is even stronger when there is a difference to be accepted and worked with.

Sexuality, animal or human, is basically biological. This means that it is regulated from the midbrain. There is, however, a chronological difference between biological sexuality and all other biological functions such as breathing, hunger, thirst, sleep, defecation, urination, etc., which are functional from birth and essential to life. Sexuality, on the other hand, is functional only at puberty at which time the hypothalamic centers start to stimulate the sexual centers (pituitary glands, ovaries, testes). Human biological sexuality appears at puberty, i.e., only after the potential for almost the whole intelligence has been activated, after a large part of human culture has been integrated. As far as human sexuality is concerned, one expects that sexual behavior be under full neocortical control, adapted to the culture and habits where one lives, and not under sole control of the mid and hindbrain centers as it is in animals. We see a lion, sea lion, or baboon defending a group of females. This is only defense of a possession, and there is no female participation or collaboration in this defense. There is no real togetherness in animals, and the basic elements of human togetherness started in a man-woman relationship, specifically during the human sexual act.

The sacrifice of the animal mother in protecting and defending her offspring has been mentioned again and again. The animal mother can be savage and courageous in defending her litter, but I have yet to see an animal mother who died defending her litter against a stronger individual. Individuals in the animal kingdom are not capable of grasping the concept of total sacrifice for another individual. The only time I have seen such sacrifice is in dogs defending a human being to whom they are attached. But that is another matter, as it involves human interference.

The concept of ultimate relationships, what the philosopher Martin Buber called the "I and Thou" relationship (a relationship between the souls) is solely a human concept and appeared somewhere during the hominid evolution. There is no doubt that it appeared in male/female relationships during the hominid evolution. It is only when the concept of love appeared between the hominid male and female (started during the sexual encounter) that the idea of ultimate sacrifice appeared. I would not hesitate to state that the concept of ultimate sacrifice appeared first in Henrietta in love and then in Henry in love. That a weak female would stand up to a strong male is something completely new in the mammalian world. It extended to males, generalized to progeny, siblings, the whole group.

In humans, women are interested in sex, and this does not exist in animal females. The female of the human species discovered sex and wanted to become sexually attractive to participate in it. Having lost pheromones, Henrietta wanted to create a new type of sexual stimulation, which would make her not only sexually attractive to males but attractive in general. She wanted to become sexually attractive because sex had become pleasurable to her. Subsequently, Henrietta wished for and looked forward to sex, and Henry easily followed.

Taboos and Rituals

Finally, men and women needed to organize and discipline the practice of sex, rid the haphazard and disorganized practice used by animals, and control male aggression. A method to control aggression and biological sexual drives is through the practice of taboos and rituals. These practices regulate any sexual behavior; namely, aggression, that could become harmful to women or children.

Taboos and rituals make the new sex official and consecrated. It became second nature, as natural as the one it just replaced.

Concepts of Space and Time

The concepts of time and space barely exist in animals. The animal knows only that something is or is not happening, is or is not there. Animals barely know if something will or has happened, or is near or far away. This is what is meant when we say animals barely have any notion of space and time.

Humans have developed the concept of space and time. They know that an event has happened or will happen, they remember the past and know how to anticipate and wait. They also know what is far away and what is close. They even have a combined space-time concept since they know how to relate a distance to the time necessary to cover that distance (Hawking, 1988).

The concepts of space and time appeared somewhere during the hominid evolution, when there was a need for these concepts to appear. They appeared when the fear of bad memories, satisfaction of good memories, or anticipation or fear of a coming event were perceived. They also appeared when an event was associated with a specific location that is close or far away. Human sexuality played an important role in the acquisition of the time and space concepts because they became necessary for adequate functioning of human relationships. The loved (or hated) person was here a long time or a short time ago; will come soon or in the distant future; the loved (or hated) one is very near or very far; the person is alive or dead, or is just a fantasy. I believe it is with the emotional aspect of sexuality that the concepts of time and space originated.

Differences Between Henry and Henrietta's Sexual Evolution

I have described the evolution of human sexuality stressing problems of Henrietta's evolution. She was presented as the primary initiator, with Henry being more cautious, at least in the beginning. She took this leading role because she had much more at stake and more transformation was needed. Henry, on the other hand, did not have any urgent problem. He already owned some primitive orgasm in the form of ejaculation, and he was very comfortable with the ventral-dorsal posture. At any rate, he started to confront Henrietta face-to-face during coitus and afterward when emotions and love evolved. At the beginning, Henry was physically strong but he and Henrietta were emotionally weak. In fact, Henry and Henrietta first had to mature emotionally, and maturation is a painful process.

Henry somehow had to follow Henrietta because with the new emotional dimension sex was less the exclusive privilege of the physically strongest. Sex had been democratized and made more accessible to the entire male population because Henrietta had more say in the matter. Henry desires sex more than Henrietta and, from the sexual point of view, there are more Henrys than Henriettas. Since there are many males to choose from, Henrietta will indulge her sexual favors not with the old generation of aggressive males but with a new generation of males more likely to conform to her new requirements. Natural selection quickly selected males who conformed to the new conditions.

Henry, obviously, has his own evolution which is separate and not always parallel to that of Henrietta. It would be a mistake to think that Henrietta dominated Henry through her sexual favors when both started this three-million-year voyage. Nothing is further from the truth. Henry is physically stronger, aggressive, and, most of all, he is the hunter. While Henrietta can gather fruits, Henry can provide the primary food (meat) by hunting, with which he can request favors (sexual or otherwise) or impose his will. Henry's secondary role in the sexual revolution does not mean a matriarchal society; in many present cultures, men still have the leading role. In some contemporary cultures such as the North American Indians of previous generations or certain Moslems, the status of women was or is very low. Women of all cultures are still struggling for equality.

Alterations in the hormonal cycle of Henry during the sexual evolution were by no means as drastic as in human females. The only difference is that in animal males the pituitary gland is directly activated and stimulated by pheromones produced by the female in heat and sniffed by the male; in the human male the pituitary gland receives direct stimulation from the neocortex which itself is receiving tactile, visual, auditory, and intrapsychic input of a sexual nature. The emotional nature of women manifested through sexuality was seen as an invasion of his authority and physical power and sometimes he reacted violently and aggressively to these changes. For him, anteriorization is already here; hiding his genital organs is minimally necessary. Superiorization makes less sense since originally Henry had no reason to hide his genital organs. As far as beauty and charm were concerned, he had a primitively limited interest.

As far as men are concerned, the only great advantage created by the departure of pheromones is that they are no longer subjected to a frenzy which causes them to completely lose control. Freedom from pheromones was the male's chief gain during the sexual revolution because they became independent of the practice of sniffing the female perigenital area. Henry's main involvement in this sexual revolution was represented by emotions and love and in this he was reluctantly involved, differently from Henrietta.

In comparing male and female human sexuality, the biological urge in males is still strong because, based on anatomical and physiological considerations such as distension of the internal genital organs by sperm and sperm products and the need to evacuate them, the biological urge in the female is very discrete, concealed, barely or nonperceptible, and when perceptible, it is only at the time of ovulation. The mental urge in the male is also very strong because of continuous stimulation, while the female is subjected to less sexual stimulation and is more concerned with stimulating than being stimulated.

The biological drive is still present in the male in that he wants to empty his painfully distended seminal vesicles into any vaginal cavity. It is likely that in the beginning of the hominid evolution males were forcing females to the animal posture (ventral-dorsal). However, reverting to quadrupedal posture, even if just for sex, was becoming more impractical because quadrupedalism was being relinquished, and females were becoming less cooperative (see film, "The Quest of Fire"). Therefore, the early hominid male cooperated but always missed the old ways, to which he occasionally reverted. The male's reaction against this new female behavior is to revert to physical strength, until desire for union and love took over.

Erection of the mammalian penis can take three forms depending on the species. The first form is the baculum or penis bone. These animals usually have a small external penis kept permanently erect by a penis bone. They have only a limited erection, and the rigidity of the penis during copulation is secured by the baculum. Most mammals such as dogs and most primates have a baculum.

The second form is the retractable penis. These animals also may have a baculum but their penis is permanently of large size and, most important, it is always retracted within the pelvic cavity except during urination and copulation. These animals also have a limited erection or rather their penis is permanently erect. Typical examples are the horse and the elephant.

Most mammals have some kind of erection or enlargement and hardening of the penis as a result of sexual stimulation. The degree of erection, or size ratio between the penis when sexually stimulated and at rest varies greatly from species to species. *Homo sapiens* has the greatest degree of erection. The reason may be that if a penis were more erect at rest it would be easily exposed to trauma because of erect posture. Biologists explain the length of the human penis in erection in terms of the depth of the human vaginal cavity and the necessity to deposit the sperm at the bottom of the vaginal cavity near the neck of the uterus.

It is easy to imagine how the concern for trauma and the fear of castration originated when the early hominids (australopithecines) stood up and the male organs were obviously displayed, especially during erection. All animal primates have a natural curiosity to touch any foreign object. When Henrietta made the *dreadful discovery,* she touched the foreign object, pulled on it, and tried to detach it, especially when in erection. Henry, therefore, had to hide his genitals for fear they would be harmed.

Privacy in rectum and bladder functions might have been initiated by Henrietta, who originally started hiding her anus and vagina, and Henry followed. He also had to overcome his fear of emotions and love which automatically implied giving up physical superiority.

As evolution progressed, each gender used whatever was anatomically or physically available to affirm their sexuality (the breasts and beauty for women; the penis, testes, hair, and physical strength for men). As the sexes progressively became different, two different psychologies expanded, and the two masculine and feminine personalities started being formed. They now will have to develop different approaches for their union.

Being aggressive, Henry often reacted negatively rather than followed Henrietta. Instead of submitting to Henrietta's beauty and charm and following all the stages of anteriorization, superiorization, facialization, generalization, individualization, and love, Henry occasionally reverted to his primitive brutality by going straight to her genital organs, disregarding any preparatory stage; in other words, he raped her. For our purposes, the definition of rape is simple. It is aggressive direct access to the female genital area, and regressive because it disregards all the evolutionary steps of the hominid evolution.

The remnants of sexual animality in man are to reach orgasm as often and as quickly as possible, to have many different partners, no preparation, no after sex, immediate separation after coitus, and no regard for need of female orgasm. For Henrietta, the remnants of sexual animality are sexual indifference or frigidity, fear, and occasional aggression.

Control of Aggression and Emotions

Henry's cautious approach in the sexual evolution is based, therefore, on two fears. The first is fear of castration and the second is fear of being completely seduced by Henrietta and thus being obliged to give up the superior aspect of his physical strength.

Humans would have a tendency to be aggressive from birth except for the interference of education and love. Aggression is controlled from birth, but only progressively and not always completely. At puberty there is renewed aggression in animals with remnants shown also in human puberty. In humans, personality formation and control of instincts and even sexual education are well advanced by the time puberty arrives. This makes control of sexual instinct easier, at least most of the time.

When a behavior is biological as in animals, it is determined and strictly controlled by a neurohormonal stimulation or reflex mechanism, a stereotyped behavior that is not under superior brain control. For instance, eating in animals is determined by a biological stimulation that initiates the hunger mechanism. If hunger is satisfied, there is no biological stimulation and the animal does not eat. The same applies to sex, which is solely determined by a neurohormonal mechanism (pheromones, sexual hormones, etc.). Since sexual activity is purely biological and neurohormonal with no cortical involvement, there is no abuse of sex in animals; it is biochemically controlled.

In humans, the center of sexuality is no longer in the midbrain, it is the neocortical activity (conscious or unconscious) that takes over sexual control. Sex goes from biological to essentially cerebral control. Biological sex and the aggression that goes with it can be controlled only by neocortical interference coming from the higher brain centers. Therefore, control of aggression progresses with encephalization.

The Coolidge Effect

President and Mrs. Coolidge once visited a chicken farm whose owner showed them a rooster who successively copulated with ten different hens in one day. Subsequently Mrs. Coolidge teased the President about the rooster's performance, and the President teased back by saying that it was possible only with *different* hens. Neither the farmer, President and Mrs. Coolidge, nor the humorist who subsequently called this ability the "Coolidge effect" had access to our present knowledge of sexual hormones, and all somehow "admired" the sexual prowess of the rooster. Actually, they were all practicing reverse evolution: for a human, such a performance certainly would be considered prowess, so why not for the rooster?

The truth is that the biological mechanisms regulating sexuality in humans are completely different from those in animals. To compare animal and human sexualities in terms of "prowess" is to compare apples and pears because they have nothing in common. If one is looking for similarity, I would propose that pheromones have the same effect on the male genital organs as a cathartic on the colon or a diuretic on urine formation: one produces diarrhea with the need for frequent defecation and the other a bladder irritation with the need for frequent urination. I do not see any prowess in the animal ejaculating frequently any more than in a human defecating or urinating frequently.

Looking at it retrospectively, the "Coolidge effect" is nothing more than a temporary hypersecretory irritation of the male genital organs, produced by pheromone stimulation and resulting in the need for frequent evacuation. This evacuation is done in a vagina due to scent and taste attraction, but each time in a different vagina because of an olfactory (and possibly taste) distraction from a vagina already filled with sperm and closed (Nova, Public Broadcasting System, 7/21/92, "What Smells -- a Gene Related to Immune System"). This mechanism is purely biological and in no way involves the cerebral cortex of the animal, except in how to carry out the sexual act. The animal male does not decide whether to have sex or with which partner, since these decisions are made biologically. With minor variations, this mechanism is generally the same in the animal kingdom regardless of whether one is dealing with insects, lions, or nonhuman higher primates. Finally, the original observers of the "Coolidge effect" did not record for the benefit of history what happened to the other roosters who, except the strongest one, are sexually deprived.

I do not believe that animals and humans should be compared, and I strongly believe that each should be respected and admired in their respective field. I, therefore, object to the occasional statement that animals are superior to humans, specifically in the field of sexuality, and I reiterate here my arguments. The loss of the Coolidge effect in humans has numerous implications among which are that copulations-ejaculations no longer have to be that frequent. Sexual intercourse in humans is less frequent (average two to three times a week), but it is much more prolonged than in animals. Nothing is really lost with the loss of the Coolidge effect, the sexual act is just more perfected in humans than in animals. Humans have nothing to envy in animals. The human sexual act, because it is now individualized, creates a new bonding between the two sexual partners. I have shown the birth of monogamy, fidelity, emotional involvement, love, responsibility, and also jealousy, hate, and so on. Changing sexual partners, which works against individualization, will be given up more and more (by women more than men) because humans have evolved more during the hominid evolution and are more detached from animal origins. The Coolidge effect, in terms of mandatorily requiring multiple partners, has disappeared. The evolution of human sexuality entirely explains the appearance of monogamy.

The tendency for a woman to be sexually involved is a concomitant emotional involvement. Men are also looking forward to this emotional involvement but sometimes it can be frightening to them. Sexual variety is the only way for a human male to satisfy his biological need and at the same time escape emotional involvement. Some men are attracted to short-lived sexual affairs because these can be terminated as soon as they get too involved. This, however, is far from the rule and eventually men look forward to emotional involvement as much as women.

Finally, sex in humans becomes a pleasurable function. While animals have nothing to do with sex outside the mating season, humans are constantly preoccupied with it regardless of any hormonal activation. It is obvious that pleasure in sex is characteristic of humans, and there is no evidence that it is pleasurable in animals. That the rooster and hen of the Coolidge affair enjoyed copulation is a free assumption in need of demonstration. In my opinion, this is the greatest sexual difference between animals and humans.

Henry's Constant Sexual Stimulation

There are many reasons for the human male to be constantly stimulated sexually. These reasons have to do with the anatomy of his genital organs and our present culture. For the same reasons, the human female is not constantly stimulated.

The most important reason for continuous male stimulation is purely anatomical. The male organs are external in humans. This is contrary to both female genital organs which are <u>internal</u>, and animal male organs which, although <u>external</u>, are well protected in their quadrupedal posture within a "fossa" formed by the inner hind quarters and the lower portion of the abdomen so that their "exposure" is limited.

Indeed, the genital organs in the human male are well exposed to the outside because of erect posture. They are continuously rubbing against the inner thighs or against any outside object (clothing) especially when walking, which results in constant rubbing. Any movement initiates tactile penile and testicular stimulation and even erection. This is one shortcoming (or advantage) of erect posture, and animal males in their quadrupedal posture are not submitted to this constant mechanical irritation. Since their organs are easily exposed, men are more conscious of them and this readily leads to sexual stimulation. Men hold their penis during urination and, privately or not, they continuously indulge in scratching their genitals even through their clothing. Since sexual stimulation in humans is tactile in nature, one can easily conclude that the male genital organs are continuously stimulated through touch.

There is no such external display in the human female except possibly the nipple of the breasts. The clitoris is of minute size and well hidden and protected between the labia. Females have to rub their inner thighs very hard or masturbate to elicit sexual stimulation by clitoral manipulation. The vagina and the introitus are inner cavities well protected from any tactile stimulation except during coitus. The ovaries are the most inner organs protected from any contact. I always wonder what female sexuality would be if the clitoris was very large and easily stimulated, and if the vaginal cavity and the ovaries were continuously exposed to physical contact the way the penis and testicles are! Since the female organs are not externally stimulated, the woman receives less sexual stimulation than the man.

The third reason, maybe the *most important* for more frequent continuous male rather than female stimulation, is that the organ of sperm ejaculation and the organ of urination are the same in the male. Therefore, every act of urination in the male, which occurs a few times a day, is a kind of mini-sexual experience. Orgasmic pathways in the female are vaginal and anatomically different from the male. The process of urination in women in no way leads to sexual stimulation.

Other reasons for continuous stimulation of the human male are purely cultural with numerous variations. Men and women have subtle ways of stimulating each other, but on a superficial level, the appearance and behavior of the human female is to sexually stimulate the human male. This may appear to be a male chauvinistic statement that will provoke outcries from all women who will protest that their appearance and behavior is for personal satisfaction. This is not denied, but it should not be denied that such sexual stimulation on the part of women is more apparent, although it is reciprocated by men in different ways. Male appearance and behavior is directed toward female stimulation in the field of sports, financial success, excessive spending, social and political success, etc. The fact, however, is that female stimulation is more direct and males respond. The 1994 Winter Olympics were watched all over the world, especially the figure skating. The main comment made by the male audience was their appreciation of the upper inner thighs and the whole genital area exhibited by the female champions. I assume that the female audience had reciprocal comments about the male champions.

The primordial and fundamental rule of human society is control of sexual instincts by males. It is more difficult for a male because he is more easily stimulated sexually and, therefore requires more control than do women. The creation of human society would not be possible unless rules concerning sexual practices were first established. The trend was toward monogamy but, even with polygamy, it had to be determined which male would copulate with which female or females. Animals, because they cannot control their sexual instincts, cannot form a real society. They form only what Ardrey (1966) calls a social nucleus with the strongest individual at the top and the rest of the group more or less at its disposal (sexually and otherwise).

Once humans more or less started to respect the relationship between a specific male and female, then and only then did the male-female pairs get together and form the original human society. Control of sexual instinct was the original rule of human society and has never changed. How to reinforce this rule, make exceptions, and deal with those who break it always has been essential to maintaining any human society. One of the early signs of a society in the process of disintegration is a breakdown in the sexual rules. The Roman Empire started to fall and was replaced by Christianity because of the spreading practice of sexual orgies.

My twenty-five years of experience as a marriage counsellor and discussions with hundreds of patients concerning the quality of their sexual lives has led me to this simple conclusion: the male has to fully control his sex drive and should not make any sexual advance to a female until he is invited or at least feels there will be no resistance (stealing a kiss, exploring the female body, touching her breasts, exploring her genital area). One of the secrets of success of some male partners is not so much in knowing how to approach a female but which female partner to avoid.

If the present human female has sexual desires and wants to be penetrated, it is not a trend inherited from animal ancestors. This is a part of female sexuality that has been built de novo and progressed during the hominid evolution without any vestigial carryover from the animal world. Sexual attraction of the female toward the male is not animal in origin. In conclusion, sexually speaking, the female discovered the male during the hominid evolution. Eve discovered Adam, although Adam already knew Eve.

Magic and Religion

The practice of magic and religion has been traditionally a male undertaking (although females frequently participated as Priestesses) and must have been so during the hominid evolution. The explanation is simple. While Henrietta was leaning more toward emotions and love in mother-infant bonding as well as in the revolution of sexuality, it was the opposite for Henry. He was leaning toward thinking, which attempts to explain the course of events and the relationship between events. Henry needed a new strength to deal with the invading emotions. Originally, Henry was not much of a scientist so he looked to magic for explanations. For instance, success in hunting, disease, and fertility were related to the phases of the moon. The connection between events and the need to explain them became an intellectual obsession for Henry and also was his way to respond, even control, the emotional invasion by Henrietta. As a protective reaction against Henrietta's emotional invasion, Henry finds refuge in abstract thinking, and this leads to magic. This later became a fundamental characteristic of masculinity. To counterbalance this, Henrietta became more and more at ease with emotions. Of course, Henry and Henrietta both indulged in thinking or feeling with neither being mutually exclusive. This split between "thinkers" and "feelers" is purely arbitrary and simply represents a "modus operandi," a division of work, a practical approach to a meaningful union.

Promiscuity

Ideally, no two sexual performances by the same individual are identical and each sexual act will be a new experience. Promiscuity is sexual activity with multiple sexual partners. Promiscuity in animals is completely different from that in the human species in its intent and performance. When an animal male (lion or sea lion) secures a harem of females, he is not necessarily looking for variety (sexual acts with different individual females). The animal male does not care which female he starts or finishes the day with, does not have a preferred female and, for that matter, does not know if he has repeated the sexual act with the same female.

Promiscuity in humans may be a remnant of animal sexuality but it is completely different in that it specifically looks for change of sexual partner. The search for multiple sexual partners in some respect can be considered a human acquisition that appeared during the hominid evolution. Human promiscuity is more common among males than females. Even when the human female has multiple sex partners, she usually does not mix them. She becomes involved with one partner at one time, but when she starts a new relationship, she usually will terminate the previous one. Human males on the other hand, mix female partners more easily, and are more likely to be involved in more than one sexual relationship at the same time.

Human promiscuity can be explained. The biological part of the sexual act is the same with different partners but the sexual partner is also looking for individualization. Since the human sexual act is with a specific individual, who the sexual partner is becomes crucial in the sexual encounter even when promiscuous. It is more than a penis-vagina encounter, it is the encounter of two individuals. Men and women look to the sexual act for something magical, mysterious, a specific something to be found in the individual sexual partner. When it becomes obvious that there was nothing special, magical, or mysterious, the relationship ceases and a search starts for magic and mystery in another individual. The tendency toward promiscuity in humans is a continuous and hopeless search for individualization. I believe there is a more fundamental reason for human promiscuity, and it is the very avoidance of individualization. Human individuals may change sexual partners because they are fearful of getting emotionally involved. A relationship that goes beyond biological sex requires emotional maturity, and one may wish to end it before it goes too far. The only way to do this is by starting another relationship, also to be interrupted if it goes too far. Since the search for individualization and commitment is more frightening to men, promiscuity is, therefore, more frequent in men than in women.

A woman who has excessive sexual intercourse usually with multiple partners because it is difficult for a single male to provide enough coitus, is sometimes called a nymphomaniac. This implies she has multiple and repeated orgasms. This is far from the truth, and during my career I never met a woman who experienced "multiple" orgasms. A nymphomaniac is a woman who never reached orgasm, who is desperately pursuing it and trying to reach it by continuously changing sexual partners. The procedure is always counterproductive.

Prostitution

I have shown that the human female has almost completely lost her biological sexual drive. The return to biological sex for a woman is very difficult, if not impossible, and degrading since it is no longer part of a woman's nature. For a woman to prostitute herself several times each day, the sexual act has to be purely mechanical and devoid of all emotion. In exchange for this mechanical sex act, the prostitute expects payment. Individualization, so meaningful to a woman, is gone. She excludes emotion, first sexually and then generally from her life. There is such a thing as a prostituted personality.

Prostitution is essentially an attempt to mechanically perform the sexual act. I say attempt because it is really impossible to duplicate the animal act; and even if the two partners have nothing to do with each other before, during, or after sex, the prostitute is left with sexual fantasies, which animals do not have.

What characterizes prostitution and differentiates it from non-prostitution is the specific understanding that no trace of human emotion and no relationship whatsoever remains after the sexual act is terminated. This is what I call the prostitution agreement. If one of the two partners looks for the other to engage in repeat prostitution or if the prostitute emotionally remembers one of her clients or vice-versa, it is considered a breach of contract. When a prostitute has or prefers regular clients, she is not a real prostitute because she is, or may become, emotionally involved.

One may wonder why prostitution is condemned by men and women and why prostitutes are condemned more by women than by men. There are many reasons for female hostility toward prostitution. Prostitutes make sex cheap and ordinary; while women expect individualization and uniqueness, prostitution removes the mystery from sex.

The prostitute has sex for no other reason than financial. Let us picture such a woman in light of the evolutionary history of human female sexuality described so far.

- Anteriorization of the sexual act is still the general rule, but a request for any other sexual posture is usually granted providing the proper fee is paid. Superiorization to breasts, face, and general body, are minimalized and used only for sexual stimulation.

- Visual, auditory, and tactile stimulations are usually dispensed with or performed at maximum speed solely for stimulating the male and never the female, unless a whole scene is requested and enacted.

- Individualization is nil and all penises are alike.

- The *dreadful discovery* is replaced by exposure of the genital organs for immediate attraction and stimulation.

- There is no courtship before sex, and after-sex relationship is nonexistent.

- The slightest move toward a permanent relationship is a threat to the process of prostitution.

Essentially almost all that has been gained by women during the last three million years becomes null and void. For the female prostitute this condition is really more tragic: not only has she rejected all the acquisitions of the hominid evolution, and thereby is denied its emotional dimensions, she has no animal sexuality to fall back on.

I have met a few prostitutes as patients during my career. I quickly learned to just listen and never ask any questions: they are like sticks of dynamite waiting to explode at any moment. I have never met a person as hateful of the human condition as a prostitute. Many fall victim to pimps who exploit them ruthlessly in exchange for some appearance of humanity, and most of them live in a world of fantasy.

The male prostitute, the playboy or the man living in a state of continuous promiscuity, is not exposed to the same degree of human degradation as a female prostitute. After all, since he has not completely broken with animal sexuality he can easily fall back.

The greatest shortcoming for either partner in an act of prostitution is the inability to plan or to fantasize on its outcome beyond coitus. In the non-prostitution sexual act, either of the partners may perceive some human elements, something to relate to, in the other partner. All this has to be repressed because termination of the relationship must be strictly adhered to before the relationship starts. After all the male casually submits to this process, but for the female who submits several times a day, it becomes a living nightmare. Repressing emotions is the worse thing that can happen to a human being, and a prostitute does it continuously.

Connection Between Sex and Conception

How, and how not, to conceive during the sexual act is a problem females have struggled with during the hominid evolution and up to the present female *Homo sapiens*; one that women still have not been able to thoroughly master. To fully understand this problem, the evolution of conception must be completely separated from the sexual revolution.

The first step consists of acquiring knowledge that the sexual act can lead to fertilization. Phylogenically speaking, at what time the hominid species became fully knowledgeable that sex could lead to pregnancy is a subject of continuous speculation and fantasy in both nonscientific and scientific circles. However, both groups at one time or another credited all animals (from the most primitive forms to the highest nonhuman primates) with possessing the knowledge that copulation leads to fertilization and the 'reason' animals copulate is to reproduce and procreate. Yet, even today, there are primitive human societies who, in spite of their 1400-gm encephalized brain, do not have the knowledge that pregnancy is a consequence of the sexual act (Montagu, 1958).

I have personally treated people from the Third World who came to me with the most bizarre ideas about pregnancy (and even patients from the more developed nations). The story I hear most often is that a single sexual act performed early in a woman's life can lead to pregnancy years later. The connection of pregnancy to the phases of the moon or any other meteorological or environmental phenomenon is widespread in large parts of the world. As a practicing gynecologist, I have contributed to the education of many young girls (and boys) brought to me by their parents, who were too embarrassed to teach their children the sexual facts of life. Before starting these sessions, I always asked these young adolescents what they knew about sex and pregnancy. Most of them were well informed (even sexually active) and needed no introduction to sex. But for others, their ignorance and misconceptions were startling. Many of them only had been taught never to expose their genital area and, more importantly, never to let anyone (parents, siblings, or others) touch their genital area. The most amazing story was told to me by a thirteen year-old daughter of a colleague who said that if she sat without panties on a chair recently sat on by a male (father, brother, stranger) she would immediately become pregnant. I know scientists who attribute more knowledge to a fly or a worm (knowledge that they copulate in order to procreate and select the best sexual partner to secure the best progeny) than some humans demonstrate.

Most of the children and teenagers I spoke with received their sexual knowledge either by observing the genitals of siblings or relatives of the opposite sex or verbally from older boys and girls. Some of them had access to books or pamphlets on sex education, but most were left to fantasize or gossip among themselves. The amount of time spent by children talking and joking about sex is incredible and is usually the source of their information and misinformation. The point I make is that animals who are quadrupeds that barely expose their genital area to others, and that have no language or way of transmitting information on sex and pregnancy, are believed to know and understand that the sexual act leads to fertilization or that a pregnant animal 'knows' she is pregnant.

Once I treated and delivered a retarded teenage girl who became pregnant after being raped. Although her IQ was relatively low, it certainly was higher than that of an ape since she was capable of some language and understanding. Yet, all she understood was the rape was painful and brutal no different from any other brutal act she previously had been exposed to and certainly she was not aware of any sexual or reproductive implications. She knew her abdomen was enlarging and something inside was moving, but the fact that she gave birth to a newborn was never fully comprehended nor remembered. She had severe pains during labor, but she never realized that she was in the process of giving birth. The arrival of the newborn was a surprise, which she never related to the rape, pregnancy or labor, and her interest in the child was vague. Although this was my only experience with this type of patient, several of my colleagues have reported similar experiences, and review of the literature confirms these findings. If these retarded human females, with an IQ higher than apes, had no concept of the relationship between sexuality and reproduction, I cannot conceive that any animal could. My impression is that the newborn animal is born as a "surprise" and maternal behavior after birth (licking, breast feeding, mother-infant bonding, etc.) is biologically determined from the midbrain without any noticeable participation of the cerebral cortex. This is no different from the biological mechanism of sex in animals.

How the linkage of sex to pregnancy and the concept of pregnancy itself came about was progressively established during the hominid evolution, and certainly occurred in several steps.

First, the connection between the enlarged abdomen and the pregnancy/birth must have occurred early in hominid history. It was first understood that the newborn coming out of the vagina was the same as what was in the enlarged abdomen because something was moving within the abdomen and the abdomen collapsed after birth. At the same time a connection must have been established between the progressive enlargement of the abdomen and the fetus growing progressively within that abdomen, since women with less abdominal enlargement were giving birth to smaller and premature neonates. Therefore, the concept that the hominid fetus grew progressively in the maternal belly and not suddenly as a mature newborn, was slowly established during the hominid evolution.

Next, the connection between sex and pregnancy must have occurred much later at a more advanced stage of encephalization because, even presently, some primitive cultures with a cerebral capacity similar to the Neandertals do not yet possess this concept. My belief is that females were the first to establish the connection between sex and pregnancy, that they kept it a secret either individually or between themselves, and that it was privately transmitted from mother to daughter when language was possible. The first observation made by the female must have been the connection between cessation of menstruation (accompanied by nausea, vomiting, frequent urination, etc.) and pregnancy. Menstruation must have occurred early in the hominid evolution since it is occasionally observed in apes. The final connection must have been established at the end of the hominid evolution. It is obvious why the females kept the connection of sex to pregnancy a secret that was transmitted only to their daughters. Although it may have something to do with the secretive aspect of the female nature, it was also not to establish paternity as we will see later.

The recognition that the sexual act led to pregnancy presented advantages and disadvantages for Henry and Henrietta. The disadvantage was that now initiation of pregnancy was less of a mystery. It was the male who, through the powerful intromission of his penis in the vagina and sperm deposition, made the female pregnant. The male ego and masculinity, which had been overshadowed by female sexuality and maternity now were reinforced. The female now became less a goddess of fertility and procreation. Masculinity, virility, and physical strength were reinforced through the ability to sire and impregnate. The advantage was that the male, recognizing the neonate as his own procreation, would become a more efficient protector and provider.

Somewhere along their voyage Henry and Henrietta observed that when seeds were placed in the ground and watered, a plant of some kind grew. Henry was, therefore, quickly led to believe that his sperm was the seed that would result in pregnancy and birth. Since "territorial imperative" was the rule in mammals, Henry concluded that the ground (Henrietta) on which he deposited his seed (sperm) was his territory, as any land he cultivated was also his territory. That Henrietta also contributed an egg to the fertilization process is a concept still unknown to most present human cultures; even today civilized men tend to ignore it and accept the concept that their seed (sperm) has fertilized the ground (vagina).

This also led to the concept of virginity or virgin territory. No one really knew the time lapse between sperm deposition and fertilization, and even today in primitive cultures it is believed that deposited sperm in a vagina can stay alive and fertilize months or years after copulation. This is the case in some animal species where the male sperm is kept in reserve inside the female and only later used for fertilization. Therefore, the surest way to insure paternity was to have sex with an "intact" virgin. This first sexual act associated with a ruptured hymen and bleeding became the symbol of taking title to or possession of a woman (a very modern concept). This concept is still very much alive today although a virginal vagina can be penetrated without difficulty in about one third of women. This mentality obligates the young bride in many present cultures to simulate a hymenal rupture with excessive bleeding. The fact that the first sexual act is thought to leave a permanent scar or print on the female body still markedly determines the female position in some human societies.

There is no doubt that the establishment of paternity is an important factor in the creation or reinforcement of the family unit. Female promiscuity started to disappear to secure paternity, although males did not hesitate to continue securing a harem of females through the use of force. Another advantage of the establishment of paternity is monogamy, at least one-sided monogamy with the female more monogamous than the male.

It was only when feelings began, with the concept of individualization and love, that monogamy was more or less established. Even so, the practice of multiple female partners persisted for a long time but progressed to a preferred partner. Incidentally, the difference between an animal and a human harem is that in the former all females are equal in the eyes of the dominant male because he really does not know how to differentiate. In the human harem, there is always a hierarchy of preferences based on individualization.

The role of the human male as the maker can sometimes reach clownish proportions and, somehow, the human female reinforces it. Mothers always insist that the child resembles the father (Trevathan, 1987). That the role of the male in the process of procreation is minimal, if not nil, is more obvious in animals than in humans. The role of the male was introduced by the female during the hominid evolution when the need for a protector and provider increased. The male became protective of the female during pregnancy, supportive during birth, and provider during infancy, childhood, and even in adolescence (modern times). In summary, the establishment of male superiority was based on his physical superiority and on the establishment of fatherhood. These two phenomena established the male's possession of the children and their mother.

Statements such as "The animal female selects the healthiest and strongest male for mating to secure the healthiest progeny." are common in scientific literature (Fleagle, 1988). Such a statement presumes that the animal female has at her disposal some technique to evaluate the physical, emotional, and intellectual status of the mating partner; that she knows the sexual act results in gestation, birth, and motherhood; and is concerned about the general and future health of her progeny and her species. The animal female does not possess any such knowledge.

Separation of Sexuality from Reproduction

Once the partners realized that some sexual acts led to pregnancy and others did not, it immediately became important to find out which did and did not, and how to separate them. I am not sure if the problem occurred during the hominid evolution or only later after the arrival of *Homo sapiens*. Assuming the problem was a hominid and not a human problem, it must have occurred very late in the hominid evolution and it was dominated by magic. The first known goddess was one of fertility.

Darwinian evolution has perfected a system of reproduction in the animal world which is composed of different phases, all under biological regulation. One of the phases, copulation or interaction between male and female individuals, is also under complete biochemical regulation. Natural selection has perfected copulation and all aspects of animal sexuality in general solely for the purpose of reproduction. Teleologically speaking, evolution did not favor individuals in whom copulation had any other goal besides fertilization. Only the part of sexuality related to impregnation and fertilization exists in animals; there is no separation of coitus from pregnancy; coitus is pregnancy.

In human sexuality, natural selection is also directed toward copulation and ejaculation with the ultimate goal of reproduction. But, these goals are far from the only ones determining human sexuality. Radical changes have occurred that make coitus completely separate from biological reproduction.

In taking control of their sexual desires, men and women are no longer bound to the ultimate goal assigned by natural selection in animal sexuality, namely, reproduction. Although men and women have not completely separated sexuality from reproduction, they are aware of the fact that sexuality can lead to reproduction. Incidentally, because of modern science, sex and reproduction now can be almost completely separated; it is possible to practice copulation and avoid fertilization and to produce fertilization by artificial insemination without copulation. The whole concept of sexuality has been completely reorganized and is not solely limited to the act of copulation.

The use of various methods of female birth control, either on a permanent or temporary basis, has demonstrated that reproductive capacity in women has no effect whatsoever on their sexual attraction. When a man is sexually attracted to a woman, her reproductive capacities have nothing to do with it.

Concerning the permanent methods of birth control, I have personally performed hundreds of tubal ligations and hysterectomies on women of all ages. In questioning all these women in depth I do not recall a single woman who mentioned that she has become less attractive sexually speaking to her husband or sexual partner following sterilization by tubal ligation. The only changes in sexuality reported to me following a hysterectomy is when the depth of the vaginal cavity has been shortened by the surgical procedure or when the women did not take hormone replacement after removal of both ovaries and the uterus in pre-menopausal women.

I also have interviewed at least 100 male patients who underwent vasectomy and have not found one whose sexual potency has been altered or who has become sexually unattractive to his wife or to any sexual partners. When it comes to a permanent surgical procedure, a husband and wife openly discuss who, without any subsequent sexual alteration, will be sterilized.

Concerning the use of temporary methods of birth control, women using diaphragms, pills, or intrauterine loops have never reported any sexual change on their part or on the part of their sexual partner.

The average couple perform coitus about twice a week for at least 30 years during which time the woman can be fertilized. The average couple has two to three children and, consequently, only 0.1% of sexual activity is directed toward fertilization. Therefore, it is accurate to say that human sexuality is minimally, if at all, directed toward reproduction. In fact what I and all others in the medical profession have observed is just the opposite. More men and women are fearful rather than hopeful that a sexual act may have resulted in a pregnancy. If anything, with the practice of birth control, human sexuality has improved in quantity and quality.

I also disagree with the supposedly classic concept that women are more sexually active in their reproductive years. Women under forty are involved with multiple problems, the most important are raising children and building a relationship with the husband. In my observation it is during and after their forties in stable couples with a harmonious relationship when all their problems have been solved that one finds sexual activity of the highest quality.

What motivates people and leads them toward sexual activity goes far beyond reproduction or biochemical regulation. It is important to draw a significant conclusion from this brief summary: the Darwinian laws of natural selection and survival of the fittest do not hold up well for human sexuality. New qualities that challenge the laws of the jungle are appearing (emotions, love, hate). If one is to continue applying Darwinian evolution to hominid survival, one has to accept the concept that the brute (strong, thoughtless, emotionless), is not necessarily the fittest for survival although, so far, it has been the rule for more than a billion years. New dimensions for survival are born.

Parents-Infant Bonding in Humans

The human mother-infant relationship is completely new in the mammalian world and consists of many components with not one inherited from nonhumans. There are major differences in infant care between animals and humans.

The human infant has to be initiated from the moment of birth and progressively perfected into visual, auditory, tactile, and motor (manual and vocal) relationships. From birth on, it has to learn to smile and cry to communicate. If the human neonate is born under difficult conditions (prematurity), it makes the mother's care more difficult; very often the difficult, prolonged, and painful childbirth results in the mother's inability to take care of the newborn and herself. From toilet training and language to sexual education, everything has to be taught to the human infant. On the contrary, the animal infant has to learn very little and then only does so by imitating.

Mother-infant bonding in humans does not get any help from the well-regulated biochemical mechanisms that function so well in animals. There is no placentophagy or amniotic fluid consumption to produce mother-infant bonding; there is prolactin activity and milk production, but the mother-infant bonding is not necessarily tied to it; there is resumption of sexual activity shortly after giving birth, and the mother may have to care for a young infant while pregnant again.

Except for breast feeding, care also can be provided by the father. This aspect will gain importance later. Obviously, the human mother-infant bonding has to be built on different types of bonds and of superior quality since the old bonds are not strong enough or are gone.

With biochemical regulation eliminated, the problem is how the human mother is going to find the energy to care for a newborn who has so many demands. She needs a new motivation. Fortunately, during the hominid evolution Henrietta was becoming a different person. During the sexual revolution, we have seen her transformed into a person capable of emotions, individualization, one-to-one and permanent relationships and, finally, of love. Although this love relationship was primitively sexual in nature, she is now capable of expanding it to the infant. The newborn infant is an easy target for this new love function, and the infant is in dire need of it. So, Henrietta, now capable of love after her sexual experience with Henry, will freely and abundantly indulge all the love of which she is capable on the infant. The eternal question is which came first, sexual love or infant love? It is a semantic question that can be answered by saying they reinforced each other. The reason I maintain that sexual love came first is because the mother-infant bonding was initially functional at the animal level, and there was no need to change it.

The birth of love was necessary and concomitant with the birth and survival of the helpless human child. The persistence of that love makes parents protect the child even in adolescence.

When we admire the animal maternal care, consisting essentially of allowing (but not offering) breast feeding and continuous licking, we forget that with such limited care by human standards, no human infant could survive more than a few days; and assuming it survived, it would be an animal-like creature (Rymer, 1992). It is important to emphasize that the animal neonate does not need any love for survival, at least in the early stages, although elements of it are not denied. There are enough biochemical mechanisms to insure its survival without the human type of love.

Non-human primates, specifically higher primates, appear to represent a transition but they are much closer to animals than to humans. Placentophagy, amniotic fluid consumption, and lactation still play a role in mother-offspring bonding, but the role of touch, sight, and limited hearing also can be seen in their bonding.

The helplessness of the human female in late pregnancy, labor, and after delivery, in providing food for herself and her infant, makes her selective in choosing a sexual partner. He is constantly needed to provide food and protection. In a way she created paternity before the male became aware of it. Today, single mothers frequently give the impression that fathers are useless. The truth is that our system of welfare and medical care is diminishing and even replacing the father's role as provider.

Fatherhood concerns any kind of role played by the father in providing food and shelter to the offspring. Fatherhood is essentially limited to humans, although one can cite isolated exceptions in different mammalians. In nonhuman primates, parent-offspring caring and bonding is left exclusively to the female with minor exceptions, and the only male role is when a high-ranking male takes a protective attitude toward all the females (including children, but only secondarily). When there is a change of hierarchy with a new high-ranking male the outcome for infant and juveniles can be frightening since they can be expelled and even killed (Hrdy, 1995).

The transformation from the animal mother-offspring bonding to the human one and emergence of father-offspring bonding involved more feelings and intellectual activity. It is obvious that although maternal bonding has been transformed, it is more active than paternal bonding which, at least in the beginning, is less intense, more fragile, and easily broken because it is not reinforced by natural elements such as lactation and physical contact.

Incest

There are numerous differences between the animal and the human family unit. One outstanding feature is that the animal unit (composed essentially of mother and offspring) is disrupted as soon as the offspring shows signs of puberty or as soon as the mother stops nursing and resumes sexual activity. The human family unit, which almost always includes the father, is not disrupted as soon as the mother resumes sexual activity or stops nursing, or at puberty of the offspring. During adolescence, which is between puberty and adulthood, the human offspring is still part of the family unit. It is easy to understand why the human offspring at puberty is still part of the family. Henrietta and Henry are not going to reject Henry junior or Henrietta junior just because some pubic or axillary hair appeared, or the penis of Henry junior shows some tendency to erection, or the breasts of Henrietta junior are burgeoning. The biochemical bonding has been deactivated, and a new bonding of emotions, love, and all kinds of different attachments have appeared. The parents are not going to reject these young adolescents because they know too well that their children are not ready to go on their own and they are still attached to them.

But once Henrietta and Henry, somewhere during the hominid evolution, decided to keep Henrietta and/or Henry junior at home for a few more years until these adolescents felt strong enough to be on their own, a serious family problem became obvious. Henry and Henrietta are sexually active, and it is easy to imagine how much these adolescents were exposed to the sexual activity of their parents either through sight or sound. Crisscrossing of sexual attraction is bound to explode between Henrietta junior and Henry, Henry junior and Henrietta, or simply between the two juniors.

If Henrietta and Henry are going to stay together, either the children have to go before they are ready, or some strict rules must be established to keep the family together. How much time was spent in searching, testing, or reinforcing all kinds of rules is hard to say, but there was almost three million years to find the proper method. Before a standard was established, there must have been a lot of rejection, anger, explosions, and even frequent killings. Finally, the following two rules were established:

- Sexual activity between parents and their children is taboo.
- Sexual activity between siblings is taboo.

These two rules successfully kept the family unit intact and after being reinforced for millions of years it became almost genetic and functioned as a second nature. They certainly are the first taboos to appear in human history and have never changed. Incest is still a horror today. Additional observations and rules served to reinforce the incest taboo, as follows.

Sexual activity between Henry and Henrietta was not to be performed in front of the children so they would not get any ideas. Anyway, individualization of the sexual act called for privacy.

Childhood and prepuberty were the best times to establish conscious control on progressive biological demands. The human adolescent was, therefore, prepared and conditioned during this prepubertal time to control any biological instincts even before the push by any sexual hormones. When sexual hormones appeared at the beginning of puberty, there was bound to be some interaction and even conflict between conscious control and the sexual push resulting from the sexual hormones activated at puberty. If one becomes sexually active at puberty, one does not learn to control it because it has to be immediately satisfied, as in animals; this tends to become a life-long pattern. Sexless adolescence is a human acquisition and makes us different from animals.

Henrietta and Henry, a new type of concerned parents, were quick to recognize that sexually active young adolescents were difficult or impossible to control, and certainly did not reinforce elements of a family unit. Therefore, for the first few years of puberty and until such time as conscious control could be fully established, sexual activity in young adolescents was strongly discouraged even with sexual partners outside the family unit. Control of sexuality at the time of puberty is one of the greatest acquisitions of human society.

Pregnancy and childbirth in young teenagers were dangerous and even deadly for the young mother as well as for the neonate because there was not enough space to grow in the small body of the young teenager. Delivery was very difficult and even impossible through the small birth canal, and the very young mother was not mature enough to give proper care to a demanding neonate. As soon as the relationship between sexual activity and pregnancy was established, it became obvious that motherhood was harmful to the young teenager. Sexual activity at that young age was therefore discouraged in Henrietta junior.

Finally, even when incestuous sexual activity occurred in late adolescence or early adulthood, it became obvious that the product of incestuous conception was predisposing to severe congenital defects, namely, congenital deafness. This was the "coup de grace" that definitely condemned incest.

As the human personality was being formed during the hominid evolution and becoming more and more complex, it was obvious that biological satisfaction was not the only motivation leading to sexual activity. Individualization was playing a more important role, and feelings were becoming more involved. Besides its mechanical aspect, sexuality was becoming a confrontation, an exploration of the unknown in oneself and one's sexual partner. Therefore, there was a need to meet a stranger, someone different from the family circle. Incestuous sex is nothing more than refusal to meet the unknown, to leap into and explore the universe. In fact, this reason may be the most important in keeping the incest taboo alive today. The purpose of family education is to form and strengthen the children and the adolescent personality to approach and face the outside world without fear, and to procreate. To procreate means to create further, beyond what has been acquired in the family circle. Animals (and incest) do not procreate, they just duplicate.

When a man and a woman are willing to face the universe hand-in-hand, they have to leave and separate from their respective families. They are not rejecting their past since their past education makes them what they are today, but they are not to duplicate their past either. It is in their sexual life that Henrietta and Henry envisioned the future of hominid evolution and it is in their respective sexual life that Henrietta junior and Henry junior will move forward in the hominid evolution. Incest freezes the children in a family circle and makes them unable to separate themselves and move forward. Incest is nothing more than a regressive duplication and a denial of procreation.

All the historical reasons that led to the appearance and persistence of the incest taboo during the hominid evolution are present and alive today with similar ill effects as consequences. If the factors that led to the incest taboo had really disappeared and been lost in history the way many scientists proclaim (Montagu, 1971), the incest taboo would have disappeared or at least become attenuated. The very fact that incest is still strongly alive today is proof it is still conditioned by the same factors that originated it.

To fully understand the impact of the incest taboo in our present culture, let us imagine how our present family unit would function with no incest taboo. Such a speculation is not pure fantasy because it is not unusual to see some "advanced" elements of our present society claiming that incest could be dispensed with because it is an old-fashioned and useless taboo (Montagu, 1971). If this was the case, the first sexual encounter of the young male adolescent in the process of puberty would be with the mother because she is already there and a maternal attachment exists. The mother may sexually prefer her son; father and son will hate and kill each other; the son will never control his sex drives and will grow up with tremendous aggression. In fact, he will not grow but will remain atavistically attached to the mother.

A similar scenario will be repeated between a father and a growing daughter. Two siblings of different sexes also will be sexually involved, will unite against parents, and the family will become explosive. The family unit would be completely disrupted or rather the present family unit would never have appeared. In fact, the human culture would not have appeared at all in its present form if it was not for the incest taboo. The present supposition of free incest is not a fantasy because, as demonstrated by Freud in his study of the Oedipean complex, a tendency to incest is still present, always ready to erupt, and always has to be contained. In conclusion, the incest taboo not only contributed to the appearance of the present human culture but is also necessary for its continuance.

The act of parenthood, specifically motherhood, consists in

keeping the family together and the love relationship intact, while eliminating any sexual connotation between parents and children and between siblings. This is easier said than done. Rules concerning exposure of breasts, genital organs, unusual nudity, are strictly enforced in the family circle, especially when puberty is about to occur. Functions of urination, defecation, and sexual activity between parents become more and more private. Even sexual discussions between members of the family are more and more restricted although they are discussed more freely outside the family. Cultural variations are many, and in some cultures any kind of communication between a parent and a child of the opposite sex and between siblings of opposite sex ceases at puberty (Haley, 19). The incest taboo is not necessarily limited to the relationship between parents and their genetic children, although it is strongest in that relationship. It can involve an extended family living under the same roof regardless of their genetic relationship. In some cultures it can extend to the whole village, and spouses have to be acquired from outside the village. All these rules have a tendency to be broken in different degrees and need continuing reinforcement.

Cannibalism

Cannibalism is eating the flesh of one's own species. It is commonly observed among nonhuman mammals, including nonhuman primates (Haley, 1976). The biological attachment of the animal mother for its litter could easily reverse to cannibalism of the litter if lactation ceases and the mother is hungry. In fact, it has been demonstrated in some species that pheromones produced by the litter and sniffed by the mother is what prevents her from eating the litter. This phenomenon may be general in the animal world. Among carnivores, a constant concern of the mother is to protect her litter from predators, and this includes her own species and even the father of the litter, assuming he can be traced. Cannibalism may be more common than reported by biologists. Starvation is a way of life in the jungle and, therefore, animal cannibalism is more frequently encountered there.

Anthropophagy or cannibalism in humans is a very strong taboo. When encountered in some present primitive societies it is more a matter of ritual (eating human fat, drinking human blood) than one of nourishment. Even in these primitive societies it is not as frequent as commonly believed. Lately there has been a resurgence of interest in human cannibalism as evidenced in numerous publications, films, and even stage plays (New York Post, 5/12/92). The appearance of the cannibalism taboo in human evolution certainly preceded the incest taboo, but the mechanism of its acquisition is most likely similar. Humans could no longer rely on pheromones to repulse each other, cannibalistically. But the need to form a stable family unit and a social nucleus primitively required this taboo against cannibalism.

The emergence of thinking, feeling, emotional attachment, love, even in the most primitive form, completely ruled out cannibalism. The new mother, in her hominid and early human forms, has developed an emotional attachment to her offspring and is in no need of the biological repulsion by pheromones (which has disappeared) for not eating her infant, even if she is no longer lactating. Today, anthropophagy is seen only in the case of severe starvation combined with extreme isolation from any human society, or in extreme pathological conditions. Anyway, strong respect for the dead body in our present culture comes from the belief in the existence of the soul: this soul may be around the dead body and one would never interfere by anthropophagy.

Prominence of Henry's Role in the Family Unit

Henry is no longer controlled by pheromones, has at least managed to control his aggression, has lost the Coolidge effect, can control the animal tendency to promiscuity, is aware of his role in female fertilization, and of the incest taboo. These changes are relatively moderate compared with the tremendous upheaval of Henrietta's sexual evolution. Henrietta was essentially the leader in this revolution with Henry following most reluctantly. One wonders why today we are not living in a matriarchal society with women leading in every respect and men playing a subordinate role.

Even today there are still some primitive societies, mostly in the Pacific Islands, that are matriarchal, with women freely selecting and rejecting sexual partners and also making important decisions in the family unit (Mead, 1963). It is also true that these societies are disappearing and just must be remnants of prehistoric societies. Most of our present societies are definitely patriarchal, although most women are fighting for equality.

In some respects, male superiority is and has been more perceived than real. Women often have subtle ways of imposing their will since men do not want or are afraid, to displease their wives or sexual partners. Although from outward appearances the man makes the decisions, one can frequently detect a feminine power present in the humblest home as well as in the offices of Heads of State. History is replete with reports of women as the real power behind the throne. This practice is certainly more widespread than it appears, because it is more effectively performed when unobtrusive. Many men will not make a decision without first consulting with their wives. However, in my judgment male superiority or whatever is left of it still is the rule in most of our present cultures. There are four main factors that have fostered this perception.

1. Man is still physically stronger. The extreme is that presently there are some cultures where wife beating is still acceptable. Without going to this extreme, the superior physical dominance may be an atavistic remnant, but there are work projects in the house that only a man can do because they require physical strength. Of course, that man need not be the husband.

2. The male is now the protector against danger and enemies and also is the food provider. These two roles belong to the mother in the animal family unit, and it must have been so in the early stages of hominid evolution.

3. The third factor was the establishment of paternity.

4. With the formation and progressive complexity of human society, it became a male society. The females were essentially occupied with their sexual revolution, raising children, and preparing food. Each woman was isolated within her family unit with rather limited communication with other women in different family units. But male activities were essentially outside the family

unit and they were able to communicate between themselves and exchange ideas, at least more so than women. Men were making the rules of human society and for their masculine roles but giving little, if any, attention to women's needs. Even when a woman reaches high achievement in our present society, more often than not she has to conform to masculine standards.

The concepts that men are more thinking than feeling, and women are the opposite, and that a leader has to be more thinking than feeling are not necessarily true and are mostly cultural.

All the traditions that so far have determined the man-woman relationship in the present society are being challenged. The relationship is being totally reconsidered, and there is a search for a new basis for a more satisfactory union of men and women.

ONTOGENY AND PHYLOGENY OF HUMAN SEXUALITY

Ontogeny of Human Sexuality

Sex in teenagers, especially younger (11-14 years old) teenagers, is in part biological with some animal characteristics because their sexual act is rapid without prolongation of any phase. They do not look for individualization and will easily change partners, with no concern for or knowledge of female orgasm. The non-participation of the female teenager in the beginning of her sexual life is well known, and her sex drive is motivated mostly by curiosity, not hormones. Teenagers look for sexual relief with minimal or no concern for each other. This has some similarity with animal, or at least prehuman, behavior.

Early in puberty, the young teenager has to learn to control the sexual drive, the same way he learned early in life to control other functions (urination, defecation, eating, reaction to pain, etc.). Learning to control the sex function differs from the other functions because it begins later in puberty. The young teenager is recreating the struggle early man faced in controlling his sex drive and satisfaction, and instead slowly substituting love.

Sex education during adolescence is an important stage preparatory to sexual maturity. For a male, it consists of becoming accustomed to a female presence and her femininity without "jumping" on her. It is important that a slow and progressive interaction develop between two adolescents of the opposite sex, that there be a tolerance for each other's presence without violence and fear, that touching and kissing can take place without sexual intercourse, and that the two need not evade each other. A slow preparation for sex during adolescence is more meaningful than a brutal sexual exposure early in adolescence, which can have dangerous repercussions later in life.

The various emergent steps of sexuality as they progressively appear in human growth from birth to adulthood, vaguely duplicate the emergent sexuality of the hominid evolution. Ontogeny is not real phylogeny for many reasons other than a difference in speed of evolution (Gould, 1979). The ontogenic changes will modify, add, and dispense with many aspects during the process. One is free to assess whether or not there is a parallel.

Contrary to animals, humans become aware of sexual anatomical differences well before puberty, since the age of the *dreadful discovery* for girls is between three and four. At the same age, boys become aware of the presence of the penis-testicle and the fear of castration.

In humans there is a prepubertal stage of human sexuality which, in my opinion, is crucial in the formation of human personality. Although there are no hormonal stimulations before puberty, boys and girls are submitted to sexual stimulations of a visual, tactile, and auditory nature during their entire childhood that slowly form their masculinity and femininity. The visual differences include clothing, hairstyle, physical activity, etc. An auditory difference is a charming or an aggressive voice. The tactile differences involve whom and what to touch. These sexual activities are purely mental and not biological. There is no real sexual interest at the childhood stage, although some pleasurable sensations can be produced by masturbation.

The important aspect of human sexuality is that puberty is anticipated and expected; and the girl or boy is not unprepared. Specifically, the girl is waiting for the appearance of pubic and axillary hair, breast formation, and menstruation that confirm her sexuality (New York Post cartoon). The pubic hair is a way to hide and prepare for "the mystery." Superiorization begins as soon as breasts start to appear and increase in size; facialization expands, although it may have started in childhood; generalization and individualization (interest in a specific boy) follows. The idea that a mystery is hidden within their body and must be kept out of sight (learned early in childhood) is now reinforced. The concept of beauty and charm becomes a constant concern and preoccupation. Feelings of inadequacy, not being beautiful and attractive enough, alternate with ideas of possessing a mysterious and magical power. Fantasies abound. All this starts before puberty.

Boys are lagging in onset of puberty, which appears around 12 or 13 years old, compared with girls, who begin puberty around 11. Besides, they prefer to hide behind their aggression and physical superiority rather than give in to the beauty and charm displayed by the young female adolescent. In addition, they fear not being able to control their sexual drive, which is preferably accomplished very slowly and progressively. The cultural surroundings play an important role in restraining sexual fantasies of rape, incest, etc. Eventually, relative control of the sexual drive is fully established before and during biological puberty and in the latter part of the teens. The young man is capable of individualization which best controls biological drives. This ideal sequence, however, is too frequently broken by early sexual activity when the hormonal drive takes over the process of individualization and rape and incest are not uncommon. Unusual attachment to, as well as brutal separation from, their mothers is typical of maturing boys. Maturing girls cling to their mothers or compete with them.

At the opposite end of puberty, menopause and sexual activity post-menopause are almost specific to humans. Most nonhuman mammals do not experience menopause; they keep producing hormones and getting pregnant until the end of their lives. *Homo sapiens* and some other higher primates (gorillas, chimpanzees, orangutans) living in contact with humans may experience menopause; when and if these animals do, it lasts only a couple of years and any sexual activity automatically ceases. It is not so in humans; sexual activity is not interrupted as a result of menopause whether it occurs naturally (in the early fifties) or is surgically produced. Continuation of sexual activity in spite of cessation of hormonal activity is one of the greatest acquisitions of the human species. This should not be overlooked because, under modern living conditions, postmenopausal life easily represents one third of an ordinary life span and one half of our adult life span. Also, sexual hormones are not as fundamental as they are in animals.

Surgically castrated animals no longer have a sexual drive and cease all sexual activity. In humans, surgical castration (removal of both ovaries [oophorectomy] or both testes [orchidectomy]) does not interfere with sexual activity when these surgical procedures are performed in adulthood, even if sexual hormones (estrogen, testosterone) are not artificially substituted. If surgical castration is performed before adulthood however, artificial administration of sex hormones is necessary for sexual maturity of the body. This demonstrates that human sexuality is minimally dependent on hormones. I have personally dealt with many patients in advanced age (seventies and eighties), and I have come to the conclusion that as long as a man or woman is in good health, has no severe emotional stress, and a sexual partner is available, there is no age limit to sexual desire and activity (Abitbol and Davenport, 1974).

In summary, sex hormones are necessary for both sexual maturation and sexual activity in animals. In the human species, sex hormones are necessary for sexual maturation of the body but not for sexual activity in advancing age and after menopause. Further investigation of these similarities and differences are presently and actively being done.

In conclusion, sexual ontogeny may be reminiscent of sexual phylogeny, but it is not an exact duplication.

The accurate chronological sequence of the various stages in the emergence of human sexuality during the hominid evolution is essentially unknown. The successive steps in the formation of human sexuality from birth to adulthood are essentially well known. The two sequences are not necessarily parallel, and ontogeny is not necessarily an abbreviated phylogeny. The order of the steps in the two sequences is different since some steps may be emphasized in one sequence or completely bypassed in the other. However, a careful look at the human sequence could be instructive in attempting to reconstruct or at least better understand the hominid sequence. For instance, anteriorization and expansion of the visual, auditory, and tactile senses in sexuality were certainly emphasized during the hominid sequence, but not in the human one because erect posture occurs quickly and encephalization is already present. The *dreadful discovery* and the reliance on breasts, facialization, and generalization must have points of similarity in both sequences, and one suspects that the order of appearance was similar. Burials, art, religion, visible late in the hominid evolution are reminders of similar concerns about life and death as seen in late adolescence. Further research will uncover more similarities as well as more differences.

The Concept of Phylogeny

I have presented the emergence of human sexuality analytically and not chronologically. I have described various steps in the formation of human sexuality but not the chronological sequence, knowing well that some of the steps described first might have occurred later in the hominid evolution and vice-versa. The methods used for this chronological reconstruction of sexual events involve various scientific disciplines.

Anthropology

The australopithecines, Lucy and STS 14, lived 2 and 3 million years ago. They stood erect, but their brains were no bigger than chimpanzee ones.

Homo erectus lived between 1 to 1.5 million years ago. The size of their brains were between those of chimpanzees and modern humans.

The Neandertal lived between 50,000 to 150,000 years ago. Their brain size is similar to ours.

Afterward, starting 50,000 to 80,000 years ago, we dealt with Homo sapiens first in the archaic and then in the modern form.

The two main anthropological events during the hominid evolution were erect posture and encephalization, in that sequence. Each one played an important role in sexual evolution.

Erect posture contributed to anteriorization of the sexual act, which was the initial chronological event in the sexual revolution. It was the face-to-face and ventral-ventral postures that initiated the sexual sequence. The more perfected erect posture became, the more ventral-ventral sexual posture became mandatory. Erect posture was not only a mechanical posture, it was also a way of life. It supposed full exposure of a vulnerable, unprotected anterior body. Erect posture provided manipulation and dexterity because the front limbs were now completely free of any postural functions, but it was unstable because equilibrium on two limbs was less steady than on four limbs. The consequence was that erect posture required trust; the loss of prominent canine teeth was also helpful. Full confidence had to be established in a fellow biped not to jump on you while unprotected, go after your jugular, tear your rib cage, and open up your belly.

The question is how advanced was erect posture during the australopithecine era? The consensus is that it was not perfect and, therefore, there was always the possibility of temporarily reverting to a quadrupedal posture necessary for ventral-dorsal sexuality, especially since Henry was not willing to give it up. Ventral-ventral sex was definitely established with Homo erectus when erect posture was perfected.

There is another reason to postulate that ventral-ventral sex was not fully practiced with the earliest australopithecines (Lucy). It is because Lucy was not able to take a perfectly supine posture. Since full anteriorization of the sexual act supposes easy practice of supine posture, we can postulate that although already started with the australopithecines, it may not have been completed until the end of the australopithecine era. Both imperfect erect posture and incomplete supine posture are two reasons for suspecting that anteriorization was not fully practiced by the australopithecines.

The main consequence of anteriorization of the sexual act is the expansion of sexual perception by vision, hearing, and touch. The reception and interpretation of these expanded perceptions supposes that the corresponding areas in the brain have enlarged, consequently encephalization is already in progress. No cause-effect relationship is implied, i.e., need to increase visual, auditory, or tactical functions is responsible for those specific areas of encephalization or that encephalization is responsible for those expanded functions. What is certain is that they occurred together.

There was little encephalization during the australopithecine era, and although visual, auditory, and tactile perceptions were present, cerebral interpretations were limited and, therefore, pheromones were still active. The change in sexual stimulation from biological pheromones to other sensory modes started with the onset of encephalization. The *dreadful discovery* must have occurred only when encephalization progressed.

Further encephalization, the mastery of fire, and use of primitive tools have been demonstrated in *Homo habilis* and *Homo erectus*. They also buried their dead, which supposes not only the perception of a dead person as an individual (individualization), but some emotional relationship (not yet love) with the dead individual. Burial also implies the notion of life after death, and therefore perception of time. There is no reason not to apply all these elements to sexual activity or rather to man-woman relationships, since **Homo abilis** already belongs to the genus Homo. Therefore, with **Homo abilis** (maybe earlier, but we have no archeological proof) we can imply, first, individualization and, second, beginning of emotional relationships in the sexual act and love relationships. With the notion of time being acquired, persistence and permanence of man-woman relationship is implied.

One could argue that long before the appearance of hominids, there are persistent relationships in the animal world. A group of baboons, a herd of buffalo and mother-infant bonding in tigers, all could be called permanent or, at least, persistent relationships. Animal grouping, however, is based on gregarious instinct and strict hierarchy determined by physical power. Although hominid and human grouping can be based on hierarchy and fear, there is no doubt that a new element appeared in the hominid relationship with early evidence demonstrated in the genus *Homo*. Even the biological mother-infant bonding of animals has started to collapse in humans and in many hominids. In summary, with *Homo habilis*, we have archeological findings implying individualization, emotions, and maybe even love in relationships, and most likely between man and woman.

In the Neandertal and archaic *Homo sapiens*, encephalization is now complete and the neonate is completely dependent on parents (mother) for many of its early years. One can suppose that Henry and Henrietta are nearing the end of their journey. They both have the intellectual and most likely the emotional capacity to grasp all aspects of modern human sexuality. This does not mean that the cave man and the man of modern Western civilization are alike. Tremendous technological differences separate them, and technology determines ethics. As new tools were mastered and life became easier, Henrietta and Henry could indulge in activities not necessarily related to finding food, shelter, and protection against cold or predators. They had some spare time to look at each other and themselves. Thinking and meditating were not unusual. Speech and religion may have appeared before the Neandertal, but they are certainly full blown with them. Art (mostly religious) appears on cave walls and in primitive sculptures (French cavern painting shows man and woman struggling with connection and separation).

History: Past and Present

It is my conviction that the role of human sexuality in history, past and present, has been ignored largely because of ignorance or embarrassment (Jastrow, 1984). The study of past and present history can be relevant. From an evolutionary point of view, each culture in some way represents the struggle of Henrietta and Henry during the hominid evolution.

In most of the North American Indian cultures, women were at a low social level. All the heavy and degrading work was done by them; rape was almost an accomplishment. To men the noble tasks were war and hunting. In view of the abundance of food (buffalo) and the availability of cheap labor (women), men had little to do and, therefore, they indulged in deadly wars and heroic fantasies. When the crisis with the white man came along, there was no emotional support from the women, whose support was not wanted anyway. The collapse of the North American Indian cultures and population can be explained in various ways, but one explanation I propose is that it was a war between the Indian men and the entire white culture, with less participation of the Indian women. In that war, there was minimal emotional support from the squaws; only the false pride of the male was hurt. He was humiliated when he faced the Indian female in front of whom he had been playing hero for centuries with his aggressive superiority.

Homo sapiens moved to the New World across the Bering Sea about forty thousand years ago. It can be postulated that at that time the sexual evolution in the American Indian was mostly frozen at the emotional level. I am convinced that the reason we presently see a somewhat successful attempt at survival and even rebirth of the North American Indian culture is that *now* women participate more and even have become leaders.

Another example is the position of women in the Moslem World. The female is covered from head to toe with her eyes the only means of expressing her feelings. One could conclude that the "mystery" stage is markedly emphasized here.

In the early Eskimos, promiscuity was a way of life and, therefore, individualization was limited, at least in sexuality. It is highly probable that individualization between man and woman existed in these Eskimos. It did not include the sexual relationship however, thereby weakening the individualization process; it is in sex that relationships evolve the most.

Although ancient Greek philosophers are famous for their pure and independent thinking, it is obvious from their writings that they were fascinated by women and also had a great fear of them. They found refuge in "thinking" to escape feminine takeover.

Art

Many works of classic and modern art have emphasized one of several aspects in the evolution of human sexuality. The Greek statue *Venus de Milo* is typical of female beauty generalized over the entire body with body curves artistically reproduced. *Mona Lisa* is typical of superiorization to the breasts, facialization, generalization to the hands, and an element of mystery that has never been as well reproduced since. No one has better reproduced what women hide behind their external appearance than Picasso in "Girl in Front of a Mirror", and "Girl on the Beach". "Minotaurachy", where a minotaur attempts to rape a woman, depicts the struggle of male aggression and his inability to deal with feminine beauty and charm. Marc Chagall has mastered and glorified the expression of love between man and woman in "The Lovers" and other similar paintings. George Rouault in the "Miserere" suite and in many of his other works has described love from its most primitive to its highest form. Henry Moore emphasized woman's mystery; Goya, the perfection of generalization to the whole body; Renoir, charm; Modigliani, perfection of curves, etc. These are only a few of the artistic works known to me, but the number is infinite.

Psychology

I have no way of knowing what was going on in the minds of Henrietta and Henry during the hominid evolution and what their emotions were. The problem is further complicated because each step supposes a certain amount of intelligence.

For instance, let us take the step of the *dreadful discovery.* It is highly unlikely that an animal (including higher primates such as the chimpanzee, gorilla, or orangutan) perceived the presence or absence of penis-testicles as something frightening. Some intelligence is necessary to do so, and only *Homo sapiens* with their brain development could fully perceive it. Lucy, who is Henrietta at the beginning of her journey, was not capable of perceiving the *dreadful discovery* even though she stood up. Although all non-human higher primates with brain enlargement similar to Lucy's frequently indulged in erect posture, they likely had no perception of the *dreadful discovery.* The conclusion is that the *dreadful discovery* supposes not only erect posture but a certain amount of encephalization. The human being, on the other hand, approaches the *dreadful discovery* with a fully encephalized brain and will be able to perceive immediately all its implications. Therefore, as Henrietta and Henry's brains grew, they approached the *dreadful discovery* earlier and earlier in life.

What has been said about the *dreadful discovery* also applies to other steps in the hominid sequence with each step requiring more or less encephalization. At each stage of the hominid evolution, Henrietta and Henry were repeating each step of the sexual evolution with more intellectual and emotional depth because their encephalization was progressing. Comparing the hominid to the human sequence is, therefore, basically inaccurate since the human child and adolescent are fully encephalized when undergoing theirs. Any psychological comparison between the hominid and human evolution could be misleading. Therefore, the psychological knowledge that accompanies each step of the human sequence could be informative, but only in a limited way.

Certainly further studies are needed to determine any difference or similarity between the psychological sequence during the hominid evolution (anthropological psychology) and the human child/adolescent sequence (pediatric psychology).

Psychiatry

Sigmund Freud has extensively described penis envy, fear of castration, and the incest taboo (Oedipian complex). He has distinguished the oral, anal, and phallic stages in the development of the human personality. According to Freud (1963) the development of the whole human personality is nothing more than a sexual development. We are, therefore, on common ground. Although the hominid and human sequence are not identical, there is some similarity. Evolutionary processes are elusive, subject to stagnation, complete arrest or leaps. Many psychiatric disturbances are nothing else but arrested growth and development. This brings forward a fascinating theory. Sexual deviations (there are legions) could be explained, at least some of them, as a halt in the human sexual sequence or even a regression to a previous stage. This new subject certainly can be elaborated on and sexual pathology could be considered arrested growth, regression, or emphasis of a previous hominid stage.

Sexual aggression in present male Homo sapiens is of many variations, from chronic rapists to men who are very gentle in their sexual approach to a specific woman. These varying degrees of decreasing aggression and increasing love relationships have nothing to do with hormonal levels and could resemble Henry's struggle through three million years of hominid evolution.

The psychological approach of Carl Jung (1957b) to emotional problems is based on being extrovert or introvert, on four psychological functions (thinking, feeling, intuition, sensation), and most of all on the historic acquisition of archetypes during the evolution of human culture. The two most common archetypes are repression of anima (feeling) in men and animus (thinking) in women. Most archetypes can be traced to the origin and cause of their appearance during the hominid history and also to new ones previously described. Even being introverted or extroverted can be traced to primitive forwardness or shyness to the man-woman approach. Certainly more research is warranted in this field.

That there is little or no sexuality in retarded children and adults, implies that cerebration (encephalization) is necessary for adequate human sexuality. Problems that deserve extensive investigation are sexuality in precocious and late puberty and in mentally retarded adolescents and adults. Biological puberty is present in retarded children but their sexuality is different from nonretarded ones. This can be informative in the role of biology and the normal brain in human sexuality.

Philosophy

Philosophy, as opposed to science, can propose theories of life without demonstrable facts. A scientist has the burden of proving the accuracy of his findings, but the philosopher is free to speculate. Famous philosophers have made fascinating statements concerning the evolution of men and women in the universe and some statements have repercussions in sex and love:

- Nietzche (1978): "Man is riding between his animal origin and God, on a tightrope over an abyss."
- Kierkegaard (1944): "When a man is in love with a woman, he sees in her a new woman every day."
- Teilhard de Chardin (1957), anthropologist and Catholic priest, has described the spiritual steps that man will reach in the future, starting with his present condition.
- Antoine de Saint-Exupery (1979) already had a deep perception of the third, or spiritual, dimension.
- Martin Buber in "I and Thou" (1970), as opposed to "I and It" has more often and better than anyone else, described the depth of the spiritual dimension when man and woman totally face each other.

All these geniuses of human culture emphasize the following

points: the potential of humans is yet to come; it can be realized only with better human relationships; the most important relationship to work with is between man and woman.

The truth is that evolution is still ongoing not only in time but also in space. Looking at different cultures and civilizations, or within the same civilization in different groups of society, or within the same individual, an evolution or at least great variety is seen. The sexual evolution is not finished yet, but the orthopedic evolution (erect posture) is.

End Point

The end point of human sexuality, as could be implied from this study is not always realized; far from it. It is ideal, wishful thinking. The steps are acquired differently by different men and women, by different individuals, and the variations are great. These variations are a fundamental characteristic and an important point that defines human sexuality. Most of these variations represent either an emphasis on one of the steps described or simply an arrest (refusal to go further) in this evolution. These variations are cultural or individual. Robinson (1972) states that (with erect posture) "a wholly new dimension was added to evolution, representing a major step in the evolution of evolution itself...Indeed...it is questionable whether **hominines** should be included in the classificatory system that encompasses...all other living organisms." What Robinson said about evolution in general certainly applies to sexual evolution in particular.

My final point is to explain why sexuality in animals and animal life in general is so attractive to humans. The main reason is our concern for animals and also our desire to prevent their extinction and to improve their living conditions. One should be careful, however, not to extend love for animals to rejection of the complex human condition. We descended from animals. They represent our origins from which the direction and the goals of mankind have started. Attraction to the simplicity of animal life is not a solution to fear and rejection of human values. In moving forward, there is nothing wrong in looking at the animal condition since it certainly is easier to know where we came from than where we are going.

Literature on Modern Sexuality

The end result of human evolution as far as human sexuality is concerned is well described in modern books on human sexuality. For instance, Dr. Ruth Westheimer, a well-known sex therapist, pioneered by making the topic of sex acceptable for popular discussion with lectures, on television, in the news media, etc. She is one of the leaders on the subject of modern sexuality, and author of *"Dr. Ruth's Guide for Married Lovers"* and *"Dr. Ruth's Guide to Good Sex."*

Other manuscripts, similar to those written by Dr. Ruth Westheimer, have been published by various authors. To mention some of them: *The Kinsey Report* in 1948; *Masters and Johnson* in 1960; Lonnie Berbach, *For Yourself: The Fulfillment of Female Sexuality*, New York:New American Library, 1976; Mary S. Calderon and Eric W. Johnson, *The Family Book about Sexuality*, New York, Harper and Row, 1981; Helen Singer Kaplan, *The New Sex Therapy: Active Treatment of Sexual Dysfunctions*, New York: Brunner-Mazel, 1974, and many others.

Most of them stress the "mechanics" of sex, but all of them stress the specific characteristics of human sexuality. Some of these authors briefly oppose animal to human sexuality but the differences are obvious, as described in the previous chapters. Barely anyone mentions the evolutionary history as reported in this manuscript, but all of them, without exception, stress that sexual evolution (or revolution) is still in the making, that sexuality may be biological but is also mental and emotional, that sexual impulses are to be controlled without being inhibited, that human sexuality has to be thought about and taught instead of being taken for granted, and that it has the potential of creating (and destroying) human relationships. They all, almost without exception, stress that human sexuality is completely separate from reproduction.

The origin of all these human characteristics has been investigated in depth in the previous chapters.

Summary

If I was asked to summarize the evolution of human sexuality, I would say the following.

In the tricky and torturous road of human evolution, human females were faced with males that were physically stronger, well organized among themselves, and supposedly more intelligent, or at least using the rational part of their mind more often at the expense of the emotional part. To reestablish equilibrium, women used various simple techniques during human evolution.

The first was for a woman to diminish her sexual drive by becoming less desirous for sex (at least much less than men) and to better control of her sexual drive. This was achieved by depheromonization and by "concealed" estrus. Men no longer knew if women were desirous or even receptive to any sexual advance, obliging them to court, woo, and pursue women.

The second step was to make the sexual act much more than it really is. Taking the simple act of copulation as seen in animals, women made it more fundamental and, in fact, the most powerful drive leading mankind. This was accomplished by using simple techniques that have been perfected over the last three million years, viz., anteriorization, hiding the mysterious genital area, superiorization to the breasts, facialization and beauty, and, finally, generalization of the entire female body that had become attractive from a sexual point of view. This culminated in the introduction of orgasm.

The third step is the perception of love, a transcendent feeling that moves and will move mankind further.

Although these techniques took three million years to perfect, they are quickly learned by every woman. In other words, ontogeny is a short phylogeny, and superior intelligence is not necessary to master these techniques because they have been so well perfected.

All along and more frequently in our generation, there is the continuous argument about inequality between the two genders and for access to all male prerogatives.

It would be a mistake to think that "invasion" by females is disorganizing society. They may be disorganizing a male concept of male society, but this does not mean that male society is the best. History has repeatedly shown that societies that exclude females are eventually doomed: North American Indians, Moslem cultures. Christianity spread over the world as much because of Jesus Christ as his mother, Mary. Of course, there are abuses on both sides, and the female and male genders have not reached an ideal equilibrium. The miracle of hominid and human evolution is not yet completely accomplished.

References

Abitbol MM (1989). Sacral curvature and supine posture. Am J Phys Anthropol 80:379-389.

Abitbol MM (1988a). Evolution of the ischial spines and of the pelvic floor in hominids. Am J Phys Antrhopol 75:53-67.

Abitbol MM (1988b) Effect of posture and locomotion on energy expenditure. Am J Phys Anthropol 77:191-199.

Abitbol MM (1987). Evolution of the sacrum in Hominoids. Am J Phys Anthropol 74:65-81.

Abitbol MM; Davenport JH (1974). The irradiated vagina. Obstet Gynecol. 44:249-256.

Abitbol MM, Inglis SR (1997). Role of amniotic fluid in newborn acceptance and bonding in canines. J Maternal-Fetal Investigation 6:49-52.

Abitbol MM, Udom-Rice I, Martin AK, Toledo RB, Karimi A. (1992). Fetal sound simulation using maternal voice. J Maternal-Fetal Investigation 2:233-239.

Ackerman, D (1990). A Natural History of the Senses. Random House, New York.

Allen DL; Mech LD (1963). Wolves versus Moose on Isle Royale. National Geographic (February).

Altman J (1980). Baboon Mothers and Infant. Cambridge: Harvard University Press.

Altmann, Stuart A (1962). Field study of the sociobiology of rhesus monkeys. Anna N.Y. Acad. Science 102:338-435.

Ardrey R (1966). The Territorial Imperative, New York: Atheneum Press.

Ardrey, Robert (1961). African Genesis. New York, Atheneum Press.

Aries, Philip (1982). The Hour of Our Death. Knopf/Random House, New York.

Armstrong E, Falk D (1982). In: Primate Brain Evolution: Methods and Concepts (editors). New York: Plenum Press.

Assali, NSsS (1968). Biology of gestation. Vol. II. The fetus and neonate. Academic Press, New York.

Badinter, Elizabeth (1981). Mother's Love: Myth and Reality. New York: Macmillan.

Baterman G. (1984) Carnivores. Torstar Books, Inc., New York.

Bertram, BC (1978). Pride of Lions. Charles Scribner's Sons, New York.

Bowra, CM. The Greek Experience (1959). Mentor Books.

Brazelton, TB (1979). Behavioral competence of the newborn infant. Seminars in Perinatoloty 3:35-44.

Bronson, FH (1968). In "Perspectives in Reproduction and Sexual Behavior" (M. Diamond, ed), pp. 341-361. Indiana University Press, Bloomington.

Bronson, FH (1971). Biol. Reprod. 4:344-357.

Buber M (1970). I and Thou. New York: Charles Scribner's Sons.

Caras R (1985). Mara Simba. The African Lion. Holt, Rinehart and Winston, New York.

Chusid, JG (1982). Comparative Neuroanatomy and Functional Neurology. Lange Medical Publications, Los Altos, California.

Clutton-Brock, TH (1991). The Evolution of Parental Care. Princeton: Princeton University Press.

Cornford, FM (1950). Greek Religious Thought. Beacon Press, Boston.

Creel S, Worthman C (1995). Liquid Assets: A Brief History of Wet-Nursing by Sarah Blaffer Hrdy. Natural History Magazine, (December), pg. 42. New York.

Curtiss S (1977). Genie: A Psycholinguistic Study of a Modern Day "Wild Child." Academic Press, New York.

Day, Corrine; Galef, Bennett (1995). Natural-Born Mothers, by S.B. Hrdy. Natural History Magazine (December), pg. 37. New York.

Darwin C (1859) On the Origin of the Species. J. Murray: London.

de Saint-Exupery, A (1979). The Wisdom of the Sand. Chicago: The University of Chicago Press.

De Unamuno, Miguel (1954). Tragic Sense of Life. Dover Pubications, New York.

Deutsch, Helena (1945). The Psychology of Women. Grime and Stratton, New York.

DeVore, Irven (1965). In: Primate Behavior (editor). New York: Holt, Rinehart and Winston.

Diamond J (1992). Must we shoot deer to save nature? Natural History (August).

Dollard, John, et. al (1961). Frustration and Aggression. New Haven Connecticut, Yale University Press 1939, reprinted 1961.

Dominis J; Edey M (1968). The Cat of Africa. Time-Life, New York.

Dostoevski, Fedor (1962). The House of the Dead. London: Dent; New York: Dutton.

Edey MA (1972). The Emergence of Man, The Missing Link. Time-Life Books, New York.

Ellis, Richard (1982). Dolphins and Porpoises. Alfred A. Knopf, New York.

Ewer RF (1968). Ethiology in Mammals. Plenum Press, New York.

Fisher M (1943). In Praise of Man. Charles C. Thomas, Springfield, Illinois.

Fleagle JG (1988). Primate Evolution and Adaptations. Academic Press.

Forsyth A (1966). A Natural History of Sex. New York: Charles Scribner's Sons.

Fossey D (1983). Gorillas in the Mist. New York: Houghton Mifflin Company.

Fox, GJ (1975). Pheromones in Apes. Am J Phys Anthropol 50:438.

Freud A (1936). The Ego and Mechanisms of Defense. Hogarth Press Ltd., London/

Freud S (1963). Introductory Lectures to Psychoanalysis. The Hogarth Press Limited. London.

Freud S (1930). Civilization and its Discontents. London: The Hogarth Press.

Ghiglieri M (1987). War Among the Chimps. Discover (November).

Golding W (1962). Lord of the Flies. Coward-McCann, New York.

Goodall J (1971). In the Shadow of Man. New York: Dell.

Gould SJ (1977). Ontogeny and Phylogeny. Cambridge, Mass: The Belknap of the Harvard University Press.

Gubernick DJ (1981). Parent and infant attachment in mammals. In D.J. Gubernick and P.H. Klopfer (Eds.), Parental Care in Mammals, New York: Plenum.

Haley A (1976). Roots. Doubleday.

Hassler, R; Stephan, H (1967). Evolution of the Forebrain. Plenum Press. New York.

Harlow, HF; Harlow, MK; Hansen, EW (1963). The maternal affectinal system of rhesus monkeys. In H.L. Rheingold (Ed.) Maternal Behavior in Mammals, New York, John Wiley and Sons.

Hawking SJ (1988). A Brief History of Time. New York: Bantam Books.

Hoage, RJ, Deiss WA (1996). New Worlds, New Animals: From Menagerie to Zoological Park in the Nineteenth Century, The Johns Hopkins University Press Bulletin, Maryland.

Hoogland, John (1995). Natural-Born Mothers, by S.B. Hrdy. Natural History Magazine (December), pg. 37. New York.

Hrdy, Sarah Blaffer (1995). Natural-Born Mothers. Natural History Magazine (December), pg. 30. New York.

Hrdy, Sarah Blaffer (1994). Infanticide and Parental Care, eds. Parmigiani Stefano and Frederick S. vom Saal. Switzerland: Harwood Academic Publishers.

Hrdy, Sarah Blaffer; Carter, C. Sue (1995) Hormonal Cocktails for Two. Natural History Magazine, pg. 34 (December).

Huxley A (1946). Brave New World. New York and London: Harper Bros.

Izard, MK; Vandenbergh, JG (1982a). J. Anim. Sci. 55:1160-1168.

Izard, MK; Vandenbergh, JG (1982b). J. Reprod. Fertil. 66:189-196.

Jastrow R (1984). The Essential Darwin. Little Brown and Co. Boston.

Jerison, HJ (1973). Evolution of the Brain and Intelligence. Academic Press, New York. Harcourt Brace Jovanovich.

Johnson, JI; Switzer, RC III; Kirsch, JAW (1982). Brain Behav. Evol. 20:97-117.

Jolly A (1972). Hour of birth in primates and man. Folia Primatologica 18:108-121.

Jung, C (1957a). Collected Works. Bollingen Series, Pantheon Books, Inc., New York.

Jung, C (1957b). Psychiatric Studies. Bollingen Foundation. Pantheon Books, Inc., New York.

Kant, Immanuel (1956). Critique of the Pure Reason, translated by Norman Kemp Smith, McMillan Co., London.

Kappers CUA; Huber GC; Crosby EC (1960). The Comparative Anatomy of the Nervous System Publishing Co., New York.

Keverne, EB (1974). New Sci. 63:22-24.

Kierkegaard, Soren (1944). The Concept of Dread. Princeton University Press, Princeton, New Jersy.

Kierkegaard, Soren (1941). Fear and Trembling and the Sickness unto Death. Doubleday Anchor Books, New York.

Kinsey AC (1953). Sexual Behavior in the Human Female. Philadelphia: Saunders.

Kinsey AC (1948). Sexual Behavior in the Human Male. National Book Network, Philadelphia and London.

Klaus MH; Kennell JH (1982). Parent-Infant Bonding. St. Louis, Missouri.

Klopfer PH (1971). Mother love: What turns it on? American Scientist 59:404-7.

Leakey, Louise SB (1965). Olduvai Gorge 1951-1961. Cambridge University Press, Cambridge, MA.

Leakey, LSB, Leakey, M (1964). Recent discoveries in Tanganyica. Nature 202:5-7.

Leaky, REF (1985). Homo erectus unearthed. National Geographic 168:624-629.

Leaky, RE (1986). The Illustrated Origin of the Species by Charles Darwin. Faber and Faber, London and Boston.

LeBoeuf BJ; Whiting RJ, Ganti RE (1972). Perintal behavior in northern elephant seal females and their young. Behavior 43:121-156.

Lovejoy CO, Heiple KG, Burstein AH (1973). The gait of australopithecus. Am J Phys Anthropol 38:757-80.

Martin RD (1983). Human brain evolution in an ecological context. Fifty-second James Arthur Lecture on Evolution of the Human Brain. New York: American Museum of Natural History.

Masters WH, Johnson VE (1966). Human Sexual Response. National Book Network, Philadelphia and London.

McClintock, MK (1978). Horm. Behav. 10:264-276.

McClintock, MK (1981). Am. Zool. 21:243-256.

McNiff, WT (1962). Greek and Roman Writers. The MacMillan Co., New York.

Mead, Margaret (1963). Sex and Temperament on Three Primitive Societies. Morrow Quill Paperbacks, New York.

Meredith M (1980). In "Chemical Signals in Vertebrates and Aquatic Invertebrates" (E. Müller-Schwarze and R.M. Silverstein, eds.), pp. 303-326. Plenum, New York.

Meredith M (1982). In "Olfaction and Endocrine Regulation" (W. Breipohl, ed.), pp 223-236, IRL, London.

Milton, John (1965). Paradise Lost, Books I & II. New York: Asia Publishing House.

Milton, John (1932). Paradise Regained. London: Scholartis Press.
Mohnot, SM; Sommer, Volke (1995). Natural-born Mothers, by S.B. Hrdy. Natural History Magazine (December), pg. 38. New York.
Montagu A (1971). Touching. The Human Significance of the Skin. New York: Columbia University Press.

Montagu, A (1958). Man: His First Million Years. A Mentor Book.

Moore, R (1964). Evolution. Time, Inc.

Mountfort G (1981). Saving the Tiger. Michael Joseph, London.

Nadler RD (1974). Periparturitional behavior of a primiparous lowland gorilla. Primates 15:55-73.

Newton, N (1971). Psychologic differences between breast and bottle feeding. American Journal of Clinical Nutrition 24:993-1004.

Nietzsche, Friedrich (1989). Beyond Good and Evil. Random House: New York.

Nietzsche, Friedrich (1978). Thus Spoke Zarathustra. Penquin Books: New York (c1954).

Norback, CR and Montagna, W (1970). The Primate Brain. Appleton-Century-Crofts. Meredith Corporation, New York.

Owen, E (1975). Prehistoric Animals. Octopus Books Ltd.

Oyama, T; Furusalea, T; Kito, T (1979). Color vision in man and animals. Scientific American (December).

Paglia, Camille (1990). The Sexual Personae. Yale University Press, New Haven, Connecticut.

Paton, HJ (1970). Kant's Metaphysics of Experience. Humanities Press, Inc. New York.

Pond CM (1977). The significance of lactation in the evolution of mammals. Evolution 31:177-199.

Rigler DM (1972). Attenuation of Severe Phobia. Twentieth International Congress of Psychology, Tokyo.

Robbins J (1986). Wolves Across the Border. Natural History (May).

Robinson JT (1972). Early Hominid Posture and Locomotion. Chicago: Chicago University Press.

Rosenblatt JS; Siegel HI (1981). Hysterectomy-induced maternal behavior during pregnancy in the rat. Journal of Comparative Physiology and Psychology 89:685-700.

Rubin R (1963). Maternal Touch. Nursing Outlook 11:328-331.

Rymer, Russ (1992). A Silent Childhood. The New Yorker, April 13-20.

Sarmiento EE (1987). Long bone torsion of the lower limb and its bearing upon the locomotor behavior of Australopithecines (abstract). Am J Phys Anthropol 72:250-251.

Sarnat, HB; Netsky, MG (1974). Evolution of the Nervous System. Oxford University Press, New York.

Schaller GB (1972). The Serengeti. Chicago University Press, Chicago.

Schiebinger Londa (1995). Nature's Body: Gender in the Making of Modern Science. Boston: Beacon Press.

Schmid P (1983). A reconstruction of the skeleton of Al 288-1 (Hadar) and its consequences. Folia Primatol. 40:283-306.

Schultz AH (1969). The Life of Primates. New York: Universe Books.

Shapiro H (1933). Man: 500,000 years from now. Natural History 33:582-95.

Sherfey MJ (1973). The Nature and Evolution of Female Sexuality. Vintage Books, Random House: New York.

Sinclair, Sandra (1985). How Animals See. Facts on File Publications.

Smuts B (1987). What are Friends For? Natural History (February).

St. Augustine (1960). The City of God. Desclee de Bouvier, Paris, France.

Stern JT, Sussman RL (1983). The locomotor anatomy of Australopithecus afarensis. Am J Phys Anthopol 60:279-317.

Stratton P (1982). Psychobiology of the Human Newborn. John Wiley and Sons, New York.

Tannahill R (1992). Sex in History. Scarborough House, Chelsea, Michigan.

Teilhard de Chardin, Pierre (1964). The Future of Man. Harper & Row: New York (1st American Edition).

Teilhard de Chardin, Pierre (1958). Oeuvres: 5, L'Avenir de l'Homme. Editions du Seuil. Paris, France.

Teilhard de Chardin, Pierre (1957). The Human Phenomenon. Edition du Senil, Paris, France.

Thorpe WH, North, Myles (1965). Origin of the power of vocal imitation. Nature 208:219-222.

Tinklepaugh, OL; Hartman, CG (1930). Behavioral aspects of parturition in the monkey (Macacas rhesus). Comparative Psychology 11:63-98.

Tobias PV (1971). The Brain in Hominid Evolution. Columbia University Press, New York.

Trevathan WR (1987). Human Birth: An Evolutionary Perspective. New York: Aldine de Gruyter.

Uris L (1989). The Haj. Bantam Books: New York.

Vandenbergh JD (1983). Pheromones and Reproduction in Mammals. New York: Academic Press.

Van der Post L (1961). The Lost World of the Kalahari. The Hogarth Press. London.

Waley A. (1964). Translations from the Chinese. Alfred A. Knopf, Inc. New York.

Washburn, SL; De Vore, Irven (1961). Social life of baboons. Scientific American (June).

Wexo, JB (1980). Zoobooks. Wildlife Eduction Ltd. San Diego, California.

Wilson, EO (1975). Sociobiology: The New Synthesis. Harvard University Press, Cambridge, MA.
Kant, I (2001). Metaphysics of Experience. Humanities Press, Inc., New Jesey.

McDaniel, SV (1965). The Philosophy of Nietzsche. Monarch Press,

Additional Reference Material

Abbott DH, and Hearn JP. The effects of neonatal exposure to testosterone on the development of behaviour in female marmoset monkeys. Ciba Found. Symp. 14-16:299-327,Mar. 1978.

Akers JS, and Conway CH. Female homosexual behavior in Macaca mulatta. Arch. Sex. Behav. 8:63-80,Jan. 1979.

Aron C. Mechanisms of control of the reproductive function by olfactory stimuli in female mammals. Physiol. Rev.59: 229-84, Apr. 1979.

Baggio G, and Ferrari F. The role of dopaminergic receptors in the behavioral effects induced by lisuride in male rats. Psychopharmacology 80:38-42,1983.

Banks DH, and Stabenfeldt G. Luteinizing hormone release in the cat in response to coitus on consecutive days of estrus. Biol. Reprod. 26:603-11,May 1982.

Bau, MJ, Gallagher, CD, Martin JT, and Damassa DA. Effects of testosterone, dihydrotestosterone, or estradiol administered neonatally on sexual behavior of female ferrets. Endocrinology 111:773-80,Sep. 1982.

Baum MJ, Everitt BJ, Herbert J, and Keverne EB. Hormonal basis of proceptivity and receptivity in female primates. Arch. Sex Behav. 6: 173-192,1977.

Beach FA. Animal models for human sexuality. Ciba Found. Symp. 2:113-43,1978.

Bercu BB, Brown T, and Brown H. Micro pulsatile and macro pulsatile gonadotropin secretion in male subhuman primate during sexual development. Pediatr Res. 15:1547,1981.

Beyer C, Contreras JL, Larsson K, Olmedo M, and Morali G. Patterns of motor and seminal vesicle activities during copulation in the male rat. Physiol. Behav. 23:495-500,Sep. 1982.

Bielert C, Howard-Tripp ME, Van der Walt LA. Environmental and social factors influencing seminal emission in chacma baboons (Papio ursinus). Psychoneuroendocrinology 5:287-303,Dec. 1980.

Bielert C, and van der Walt LA. Male chacma baboon (Papio ursinus) sexual arousal: mediation by visual cues from female conspecifics. Psychoneuroendocrinology 7:31-48,1982.

Bourgeois M. [Shunning of incest in human and nonhuman primates. Preliminary note]. Ann. Med. Psychol. 137:612-20,Sep. 1979.

Chambers KC, and Phoenix CH. Sexual behavior in old male rhesus monkeys: influence of familiarity and age of sexual partners. Arch. Sex Behav. 11:299-308,Aug. 1982.

Coelho AM,Jr., and Bramblett CA. Sexual dimorphism in the activity of Olive baboons (Papio cynocephalus anubis) housed in mono sexual groups. Arch. Sex. Behav. 10:79-91,Feb. 1981.

Contreras JL, and Beyer C. A polygraphic analysis of mounting and ejaculation in the New Zealand white rabbit. Physiol. Behav. 23:939-43,Nov. 1979.

Coopersmith CB, and Banks EM. Effects of olfactory cues on sexual behavior in the brown lemming, Lemmus trimucronatus. J. Comp. Psychol. 97:120-6,Jun. 1983.

Dewsbury DA, Oglesby JM, Shea SL, and Connor JL. Inbreeding and copulatory behavior in house mice: a further consideration. Behav. Genet. 9:151-63,May 1979.

Dykeman DA, Katz LS, and Foote RH. Behavioral characteristics of beef steers administered estradiol, testosterone and dihydrotestosterone. J. Anim. Sci. 55:1303-9,Dec. 1982.

Epple G, Golob NF, Cebul MS, and Smith AB III. Communication by scent in some callitrichidae primates an interdisciplinary approach. Chem. Sens. 6:377-390,1981.

Epple G. Lack of effects of castration on scent marking, displays and aggression in a South American primate (Saguinus fuscicollis). Horm-Behav. 11:139-50,Oct.11,1978.
Epple G. Studies on the nature of chemical signals in scent marks and urine of saguinus-fuscicollis callithricidae primates. J. Chem. Ecol. 4:383-394,1978.

Flannelly KJ, Blanchart RJ, Muraoke MY, and Flannelly L. Copulation increases offensive attack in male rats. Physiol. Behav.29: 381-5, Aug. 1982.

Foerg R. Reproductive behavior in Varecia variegata. Folia-Primatol 38:108-21,1982.

Fredericks CM, Anderson WR, Smith CE, and Mathur RS. Patterns of periovulatory oviductal motility and progesterone in the unanesthetized rabbit. Biol. Reprod. 27:340-50,Sep. 1982.

Gadpaille WJ. Cross-species and cross-cultural contributions to understanding homosexual activity. Arch. Gen. Psychiatry. 37:349-56,Mar. 1980.

Gallego-Fernandez A. Mechanism of action of pheromones in mammals. An. R. Acad. Nac. Med. 99:181-96,1982.

Goldfoot DA, Westerborg van Loon H, Groeneveld W, and Slob AK. Behavioral and physiological evidence of sexual climax in the female stump-tailed macaque(Macaca arctoides). Science 208: 1477-9,Jun. 1980.

Gordon TP, Rose, RM, Grady, CL, and Berstein IS. Effects of increased testosterone secretion on the behavior of adult male rhesus living in a social group. Folia-Primatol 32:149-60,1979.

Goy RW and Goldfoot DA. Neuroendocrinology: animal models and problems of human sexuality. Arch. Sex Behav. 4:405-20,1975.

Hansen S, Kohler C, Goldstein M, and Steinbusch HV> Effects of ibotenic acid-induced neuronal degeneration in the medial preoptic area and the lateral hypothalamic area on sexual behavior in the male rat. Brain.Res. 239:213-32,May 1982.

Harcourt AH. Strategies of emigration and transfer by primates, with particular reference to gorillas. Z. Tierpsychol. 48:401-20,Dec. 1978.

Harlow HF, Harlow MK, Hansen EW, and Suomi SJ. Infantile sexuality in monkeys. Arch. Sex. Behav. 2:1-7,Jun.

Hlinak Z, Madlafousek J, and Krejci I. Effects of lisuride on precopulatory and copulatory behaviour of adult male rats. Psychopharmacology 79:231-5,1983.

Irving SM, Goy RW, and Daviss GA. Copulation-induced abbreviation of estrus in the female guinea pig: block by clonidine. Pharmacol. Biochem. Behav. 12:755-9,May, 1980.

Johnston RE. Attraction to odors in hamsters: an evaluation of methods. J. Comp. Physiol-Psychol. 95:951-60,Dec. 1981.

Jonas AD, and Jonas DF. A biological basis for the Oedipus complex: an evolutionary and ethological approach. Am. J. Psychiatry 132:602-, Jun.

Kendrick KM, and Dixson AF. Ovariectomy does not abolish proceptive behavior cyclicity in the common marmoset (Callithrix jacchus). J. Endocrinol. 101:155-62,May 1984.

Kendrick KM, and Dixson AF. The effect of the ovarian cycle on the sexual behaviour of the common marmoset (Callithrix jacchus). Physio. Behav. 30:735-42,May 1983.
Kleiman DG. Parent offspring conflict and sibling competition in a monogamous primate leontopithecus-rosalia. Am. Nat.114: 753-760, 1979.

Kleiman DG, and Mack DS. Effects of age, sex, and reproductive status on scent marking frequencies in the golden lion tamarin, Leontopithecus rosalia. Folia-Primatol 33:1-14,1980.

Lemmon WB, and Allen ML. Continual sexual receptivity in the female chimpanzee (Pan troglodytes). Folia. Primatol. 30:80-8, 1978.

Mankin RW, and Mayer MS. A phenomenological model of the perceived intensity of single odorants. J. Theor. Biol. 100:123-38,Jan 7, 1983.

Marsh CW. Female transference and mate choice among Tana River red colobus. Nature 281:568-9,Oct. 1979.

McKenna JJ. Primate aggression and evolution: an overview of sociobiological and anthropological perspectives. Bull. Am. Acad. Psychiatry Law. 11:105-30,1983.

Menna-Barreto LS, Cipolla-Neto J, and Marques MD. Sexual behavior elicited in cage-mates of olfactory tubercle stimulated rats. Physiol-Behav. 31:565-7,Oct. 1983.

Michael RP, and Zumpe D. Influence of olfactory signals on the reproductive behavior of social groups of rhesus monkeys (Macaca mulatta). J. Endocrinol. 95:189-205,Nov. 1982.

Nadler RD. Reproductive physiology and behavior of gorillas. J. Reprod. Fertil. Suppl 28:79-89,1980.

Nadler, RD, Collins DC, Miller LC, and Graham CE. Menstrual cycle patterns of hormones and sexual behavior in gorillas. Horm. Behav. 17:1-17,Mar. 1983.

Nadler RD, and Miller LC. Influence of male aggression on mating of gorillas in the laboratory. Folia. Primatol 38:233-9,1982.

Nyby J. Ultrasonic vocalizations during sex behavior of male house mice (Mus musculus): a description. Behav. Neural. Biol. 39: 128-34,Sep. 1983.

Ostermeyer MC, and Elwood RW. Pup recognition in Mus musculus: parental discrimination between their own and alien young. Dev. Psychobiol. 16:75-82,Mar. 1983.

Perachio AA, Marr LD, and Alexander M. Sexual behavior in male rhesus monkeys elicited by electrical stimulation of preoptic and hypothalamic areas. Brain Res. 177:127-44, Nov. 9, 1979.

Phoenix CH. Copulation, dominance, and plasma androgen levels in adult Rhesus males born and reared in the laboratory. Arch. Sex. Behav. 9:149-67,Apr. 1980.

Phoenix CH, and Chambers KC. Sexual behavior in adult gonadectomized female pseudohermaphrodite, female, and male rhesus macaques (Macaca mulatta) treated with estradiol benzoate and testosterone proprionate. J. Comp. Physiol. Psychol. 96: 823-33, Oct. 1982.

Powers, JB, and Bergondy ML. Androgenic regulation of chemo-investigatory behaviors in male and female hamsters. Horm.Behav. 17:28-44, Mar. 1983.

Pomerantz SM, and Goy RW. Proceptive behavior of female rhesus monkeys during tests with tethered males. Horm. Behav.17:237-48, Sep. 1983.

Reynolds RL, Van-Horn RN. Exogenous estradiol and sexual receptivity in photoinhibited female Lemur catta. Physiol-Behav. 21:383-5,Sep. 1978.

R Rodriguez-Sierra JF, and Davis GA. Tolerance to the lordosis-facilitating effects of progesterone or methysergide. Neuropharmacology. 18:335-9,Apr. 1979.

Rodriguez-Sierra JF, and Terasawa E. Lesions of the preoptic area facilitate lordosis behavior in male and female guinea pigs. Brain Res. Bull. 4:513-7, Jul-Aug. 1979.

Rothe H. Some aspects of sexuality and reproduction in groups of captie marmosets (Callitbrix jacchus). Z. Tierpsychol.37:255-73, Apr.

Ruiz-de-Elvira MC, Herndon JG, and Wilson ME. Influence of estrogen-treated females on sexual behavior and male testosterone levels of a social group of rhesus monkeys during the nonbreeding season. Biol-Reprod. 26:825-34, Jun. 1982.

Ruiz-de-Elvira MC, Herndon JG, and Collins DC. Effect of estradiol-treated females on all-female groups of rhesus monkeys during the transition between the nonbreeding and breeding seasons. Folia. Primatol 41:191-203, 1983.

Rutberg AT. The evolution of monogamy in primates. J. Theor. Biol. 104:93-112, 1983.

Sassenrath EN, Jensen GD, and Blanton FL. Psychoendocrine correlates of sociosexual responsivity in older premenopausal primates. Neurobiol. Aging. 3:111-6,Summer 1982.

Smith, DG. Birth timing and social rank of adult male rhesus monkeys (Macaca mulatta). J. Med. Primatol. 10:279-83, 1981.

Stephan H, Baron G, and Frahm HD. Comparison of brain structure volumes in Insectivora and Primates II. Accessory olfactory bul. J Hirnforsch. 23:575-91, 1982.

Stoddart DM. The developing mammalian olfactory system and its role in mediating the effects of psychotropic and physiologically active odorous agents. Pharmacol-Ther. 17:65-83, 1982.

Thor DH. Gonadal testosterone and perineal sniffing behavior of the male rat. Behav-Biol. 24:256-64,Oct. 1978.

Van-Heerden J. The role of integumental glands in the social and mating behaviour of the hunting dog Lycaon pictus (Temminck, 1820) Onderstepoort J. Vet. Res. 48:19-21,Mar. 1981.

Wallen K, Goldfoot DA, and Goy RW. Peer and maternal influences on the expression of foot-clasp mounting by juvenile male rhesus monkeys. Dev. Psychobiol. 14:299-309, Jul. 1981.

Watts ES, and Gavan JA. Post natal growth of nonhuman primates the problem of the adolescent spurt. Hum. Biol. 54:53-70, 1982.

Wildt DE, Seager SW, and Chakraborty PK. Effect of copulatory stimuli on incidence of ovulation and on serum luteinizing hormone in the cat. Endocrinology 107:1212-7, 1980.

Wilson ME, Gordon TP, Blank MS and Collins DC. Timing of sexual maturity in female rhesus monkeys (Macaca mulatta) housed outdoors. J. Reprod. Fertil. 70:625-33, Mar. 1984.

Wilson ME, and Gordon TP. Sexual activity of male rhesus monkeys introduced into a heterosexual group. Am. J. Phys. Anthropol. 50:515-23, May 1979.

Wilson ME, Gordon TP, and Collins DC. Age differences in copulatory behavior and serum 17 beta-estradiol in female rhesus monkeys. Physiol. Behav. 28:733-7, Apr. 1982.

Printed in Dunstable, United Kingdom

91218262R00160